Oxidoreductase

Edited by Mahmoud Ahmed Mansour

Published in London, United Kingdom

IntechOpen

Supporting open minds since 2005

Oxidoreductase
http://dx.doi.org/10.5772/intechopen.87536
Edited by Mahmoud Ahmed Mansour

Part of IntechOpen Book Series: Biochemistry, Volume 19
Book Series Editor: Miroslav Blumenberg

Contributors
Hussein Mahdi Kareem, Luma Majeed Majeed Ahmed, Mezgebu Legesse Habte, Etsegenet Assefa Beyene, Md. Ruhul Abid, Frank W. Sellke, Maan A. Awad, Sarah R. Aldosari, Sandhya Rani Gogoi, Ana Luísa Pereira Teixeira, Mariana Gomes Morais, Francisca Guilherme Carvalho Dias, João Alexandre A. V. Velho Prior, Rui Manuel de Medeiros Melo Silva, Mrinal K. Kanti Poddar, Apala Chakraborty, Soumyabrata Banerjee, Neelima Dhingra, Ashish Bhattacharjee, Chandreyee Datta, Sukhamoy Dhabal

First published in London, United Kingdom, 2021 by IntechOpen
IntechOpen is the global imprint of INTECHOPEN LIMITED, registered in England and Wales, registration number: 11086078, 5 Princes Gate Court, London, SW7 2QJ, United Kingdom
Printed in Croatia

British Library Cataloguing-in-Publication Data
A catalogue record for this book is available from the British Library

Additional hard and PDF copies can be obtained from orders@intechopen.com

Oxidoreductase
Edited by Mahmoud Ahmed Mansour
p. cm.
Print ISBN 978-1-83880-900-3
Online ISBN 978-1-83880-901-0
eBook (PDF) ISBN 978-1-83880-902-7
ISSN 2632-0983

For EU product safety concerns:
IN TECH d.o.o., Prolaz Marije Krucifikse Kozulić 3, 51000 Rijeka, Croatia, info@intechopen.com or visit our website at intechopen.com.

IntechOpen Book Series
Biochemistry
Volume 19

Mahmoud Mansour Ph.D. is a professor of Biochemistry at the Pharmaceutical Sciences Department, College of Pharmacy King Saud bin Abdulaziz University for Health Sciences, Saudi Arabia. His specialization includes molecular biology, biochemical pharmacology, pharmacogenetics, and biochemistry. His research fields are biochemical pharmacological studies in cancers (especially hepatic cancer), antioxidants, oxidative stress, proteasome (and its role in the treatment of hepatic cancer), experimental gastroenterology, clinical gastroenterology, and diabetes. He has published more than 60 papers in peer-reviewed journals. He received the State Encouragement award from the Scientific Research Academy, (1998). Fourteen of his students were appointed as full university professors in Egypt and Saudi Arabia.

Editor of Volume 19:
Mahmoud Ahmed Mansour
Department of Pharmaceutical Sciences, College of Pharmacy,
King Saud bin Abdulaziz University for Health Sciences,
Saudi Arabia

Book Series Editor: Miroslav Blumenberg
NYU Langone Medical Center, New York, USA

Scope of the Series

Biochemistry, the study of chemical transformations occurring within living organisms, impacts all of life sciences, from molecular crystallography and genetics, to ecology, medicine and population biology. Biochemistry studies macromolecules - proteins, nucleic acids, carbohydrates and lipids –their building blocks, structures, functions and interactions. Much of biochemistry is devoted to enzymes, proteins that catalyze chemical reactions, enzyme structures, mechanisms of action and their roles within cells. Biochemistry also studies small signaling molecules, coenzymes, inhibitors, vitamins and hormones, which play roles in the life process. Biochemical experimentation, besides coopting the methods of classical chemistry, e.g., chromatography, adopted new techniques, e.g., X-ray diffraction, electron microscopy, NMR, radioisotopes, and developed sophisticated microbial genetic tools, e.g., auxotroph mutants and their revertants, fermentation etc. More recently, biochemistry embraced the 'big data' omics systems.

Initial biochemical studies have been exclusively analytic: dissecting, purifying and examining individual components of a biological system; in exemplary words of Efraim Racker, (1913 - 1991) “Don’t waste clean thinking on dirty enzymes.” Today however, biochemistry is becoming more agglomerative and comprehensive, setting out to integrate and describe fully a particular biological system. The “big data” metabolomics can define the complement of small molecules, e.g., in a soil or biofilm sample; proteomics can distinguish all the proteins comprising e.g., serum; metagenomics can identify all the genes in a complex environment e.g., bovine rumen. This Biochemistry Series will address both the current research on biomolecules, and the emerging trends with great promise.

Contents

Preface

Oxidation-reduction reactions in our body are catalyzed by a class of enzymes called oxidoreductase. The mechanism is based on the transfer of electrons from one molecule (the oxidant) to another molecule (the reductant). Oxidoreductases catalyze reactions similar to the following, $A^- + B \rightarrow A + B^-$ where A is the oxidant and B is the reductant. From a biochemistry point of view, oxidoreductase enzymes are a group of enzymes that catalyze the transfer of electrons from one molecule, the reductant, also called the electron donor, to another, the oxidant, also called the electron acceptor. Oxidoreductase enzymes utilize NADP+ or NAD+ as cofactors. Oxidoreductase enzymes include the following: oxidase, dehydrogenase, peroxidase, hydroxylase, oxygenase, and reductase. Most oxidoreductase enzymes are dehydrogenases. However, reductases are also common. The accepted nomenclature for dehydrogenases is "donor dehydrogenase", where the donor is the oxidized substrate.

Oxidases are enzymes involved when molecular oxygen acts as an acceptor of hydrogen or electrons. Whereas dehydrogenases are enzymes that oxidize a substrate by transferring hydrogen to an acceptor that is either NAD+/NADP+ or a Flavin enzyme. While the other oxidoreductases, peroxidases, are localized in peroxisomes and catalyze the reduction of hydrogen peroxide. Hydroxylases add hydroxyl groups to their substrates. Oxygenases incorporate oxygen from molecular oxygen into organic substrates. Reductases catalyze reductions and in most cases can act as oxidases.

Oxidation-reduction reactions are essential for the growth and survival of organisms. During the oxidation process of organic molecules, energy is produced. Energy-producing reactions can liberate high energy containing compounds as the synthesis of important energy molecules, such as ATP.

Oxidoreductase enzymes achieve an important role under aerobic and anaerobic metabolism. They play an important role in glycolysis, the tricarboxylic acid cycle, oxidative phosphorylation, and in amino acid metabolism. In glycolysis, the glyceraldehydes-3-phosphate dehydrogenase enzyme catalyzes the transfer of hydrogen to coenzyme NAD, leading to the reduction of NAD+ to NADH. In order to maintain the redox state of the cell, this NADH is converted to NAD+, which occurs in the oxidative phosphorylation pathway. The final pathways for complete oxidation of glucose are achieved via the TCA cycle. More NADH molecules are generated in the TCA cycle. Except for leucine and lysine, the rest of the amino acid metabolites enter the TCA cycle as intermediates of the cycle. This allows for the formation of oxaloacetate from carbon skeletons of the amino acids and subsequently into pyruvate.

Metabolic abnormalities disorders resulting from a deficiency (quantitative and qualitative) or from over-activity of oxidoreductase, which may contribute to the decreased normal performance of life, are becoming common.

Mahmoud Ahmed Mansour
Department of Pharmaceutical Sciences,
College of Pharmacy,
King Saud bin Abdulaziz University for Health Sciences,
Saudi Arabia

Section 1

Biological Application of Oxidoreductase

Chapter 1

Biological Application and Disease of Oxidoreductase Enzymes

Mezgebu Legesse Habte and Etsegenet Assefa Beyene

Abstract

In biochemistry, oxidoreductase is a large group of enzymes that are involved in redox reaction in living organisms and in the laboratory. Oxidoreductase enzymes catalyze reaction involving oxygen insertion, hydride transfer, proton extraction, and other essential steps. There are a number of metabolic pathways like glycolysis, Krebs cycle, electron transport chain and oxidative phosphorylation, drug transformation and detoxification in liver, photosynthesis in chloroplast of plants, *etc.* that require the direct involvements of oxidoreductase enzymes. In addition, degradation of old and unnecessary endogenous biomolecules is catalyzed by a family of oxidoreductase enzymes, e.g., xanthine oxidoreductase. Oxidoreductase enzymes use NAD, FAD, or NADP as a cofactor and their efficiency, specificity, good biodegradability, and being studied well make it fit well for industrial applications. In the near future, oxidoreductase may be utilized as the best biocatalyst in pharmaceutical, food processing, and other industries. Oxidoreductase play a significant role in the field of disease diagnosis, prognosis, and treatment. By analyzing the activities of enzymes and changes of certain substances in the body fluids, the number of disease conditions can be diagnosed. Disorders resulting from deficiency (quantitative and qualitative) and excess of oxidoreductase, which may contribute to the metabolic abnormalities and decreased normal performance of life, are becoming common.

Keywords: biocatalyst, biological application, disease, metabolism, mutation, oxidoreductase

1. Introduction

Oxidoreductases, which includes oxidase, oxygenase, peroxidase, dehydrogenase, and others, are enzymes that catalyze redox reaction in living organisms and in the laboratory [1]. Interestingly, oxidoreductases catalyze reaction involving oxygen insertion, hydride transfer, proton extraction, and other essential steps. The substrate that is oxidized is considered as hydrogen or electron donor, whereas the substrate that is reduced during reaction as hydrogen/electrons acceptor. Most commonly, oxidoreductase enzymes use NAD, FAD, or NADP as a cofactor [2].

 IntechOpen

Organisms use this group of enzymes for synthesis of biomolecules, degradation and removal of molecules, metabolism of exogenous molecules like drugs, and so on [3–5]. Their biochemical property such as efficiency, specificity, good biodegradability, and being studied well make it fit well for industrial purposes. As a result, oxidoreductases are being utilized in nutrition, food processing, medicine, and other chemical synthesis. In the near future, oxidoreductase may be utilized as the best biocatalyst in pharmaceutical, food processing, and other industries [6, 7].

Enzymes like oxidoreductase play great and significant function in the field of disease diagnosis, prognosis, and treatment [8]. By analyzing the activities of enzymes and changes of certain substances in the body fluids, a number of disease conditions can be diagnosed [9, 10]. The determination of the activity of the oxidoreductases is helpful in understanding the metabolic activity of different organs [8, 11]. For example, the activity of oxidoreductase enzymes in Krebs cycle is significantly increased during skin infection [12].

There are different disease conditions resulting from deficiency (quantitative and qualitative) and excess of oxidoreductase, which may contribute to the metabolic abnormalities and decreased normal performance of life [13, 14]. For example, relative decreases in the activities of NADH dehydrogenase and ubiquinol-cytochrome c oxidoreductase are highly associated with the developments of peripheral arterial disease. Another best example is mutation of p450 oxidoreductase (POR) gene, which leads to insufficiency of P450 enzymes characterized by defective steroidogenesis. Similarly, deficiency of mitochondrial acetaldehyde dehydrogenase disturbs normal metabolism of alcohol and leads to accumulation of acetaldehyde [8, 15, 16]. These conditions in turn affect the normal development and reproduction.

2. Oxidoreductase in metabolism of foodstuff

Oxidoreductases are a family of enzymes that catalyze redox reactions. Oxidoreductases catalyze the transfer of electrons from oxidant to reductant [4]. Generally, oxidoreductases catalyze reactions which are similar to $A^- + B \rightarrow A + B^-$ where A is the oxidant and B is the reductant [17]. Oxidoreductases can be oxidases where a molecular oxygen acts as an acceptor of hydrogen or electrons and dehydrogenases which are enzymes that oxidize a substrate by transferring hydrogen to an acceptor that is either $NAD^+/NADP^+$ or a flavin enzyme. Other classes are oxidoreductases enzymes, peroxidases which are localized in peroxisomes and catalyze the reduction of hydrogen peroxide. Hydroxylases are involved in the addition of hydroxyl groups to their substrates, and oxygenases are key in the incorporation of oxygen from molecular oxygen into organic substrates. And reductase enzymes are involved in the catalysis of reduction reaction [2, 3, 18]. In general, oxidoreductase enzymes play an important role in both aerobic and anaerobic metabolism. They are involved in glycolysis, TCA cycle, oxidative phosphorylation, fatty acid, and amino acid metabolism [5, 19, 20].

3. Oxidoreductase in glycolysis

In glycolysis, the enzyme glyceraldehydes-3-phosphate dehydrogenase catalyzes the reduction of NAD + to NADH. In order to maintain the redox state of the cell, this NADH must be re-oxidized to NAD+, which occurs in the oxidative phosphorylation pathway [21].

$$\text{G-3-P} + \text{Pi} + \text{NAD}^{+} \xleftrightarrow{\text{Glyceraldehyde-3-phosphate dehydrogenase}} 1,3\text{-BPG} + \text{NADH} + \text{H}^{+} \quad (1)$$

$$\text{Pyruvate} + \text{NADH} + \text{H}^{+} \xleftrightarrow{\text{lactate dehydrogenase}} \text{lactate} + \text{NAD}^{+} \quad (2)$$

4. Oxidoreductase in TCA cycle

A high number of NADH molecules are produced in the TCA cycle. The product of glycolysis, pyruvate, enters the TCA cycle in the form of acetyl-CoA. Except leucine and lysine, all twenty of the amino acids can be degraded to TCA cycle intermediates. And most of the fatty acids are oxidized into acetyl coA through beta oxidation that enter TCA cycle [19, 22].

The precursor for the TCA cycle comes from lipids and carbohydrates, both of which produce the molecule acetyl-CoA. This acetyl-CoA enters the eight-step sequence of reactions that comprise the Krebs cycle, all of which occur inside mitochondria of eukaryotic cells. TCA or Krebs cycle produces NADH and FADH, and the reactions are catalyzed by classes of oxidoreductase enzymes [23].

$$\text{Pyruvate} + \text{NAD}^{+} + \text{CoA} \xrightarrow{\text{pyruvate dehydrogenase}} \text{acetyl-CoA} + \text{NADH} + \text{H}^{+} + \text{CO}_2 \quad (3)$$

$$\text{Isocitrate} + \text{NAD}^{+} \xrightarrow{\text{isocitrate dehydrogenase}} \text{alfa-ketoglutarate} + \text{NADH} + \text{H}^{+} + \text{CO}_2 \quad (4)$$

$$\alpha\text{-ketoglutarate} + \text{NAD}^{+} + \text{CoA} \xrightarrow{\alpha\text{-ketoglutarate dehydrogenase}} \text{succinyl CoA} + \text{NADH} + \text{CO}_2 \quad (5)$$

$$\text{Succinate} + \text{FAD}^{+} \xrightarrow{\text{succinate dehydrogenase}} \text{fumarate} + \text{FADH}_2 \quad (6)$$

5. Oxidoreductase in electron transport chain and oxidative phosphorylation

Living cells use electron transport chain to transfer electrons stepwise from substrates (NADH & $FADH_2$) to a molecular oxygen. The proton gradient which is generated through electron transport chain runs downhill to drive the synthesis of ATP. Electron transport chain and oxidative phosphorylation take place in the matrix of mitochondria, and there are oxidoreductase enzymes impregnated in the inner mitochondrial membrane, which catalyze these reactions and are engaged in energy production. NADH:quinone oxidoreductase, also called NADH dehydrogenase (complex I), is responsible for the transfer of electrons from NADH to quinones, coupled with proton translocation across the membrane. Succinate:quinone oxidoreductase, or succinate dehydrogenase (complex II), is an enzyme of the Krebs cycle, which oxidizes succinate and reduces quinones, in the absence of proton translocation. Quilon:cytochrome c oxidoreductase (complex III), which transfers electrons from quinols to cytochrome c and cytochrome c:oxygen oxidoreductase, an aa3-type enzyme (complex IV), which receives these electrons and transfers it to oxygen are both oxidoreductase enzymes involved in electron transport chain and oxidative phosphorylation [19, 24, 25] (**Figure 1**).

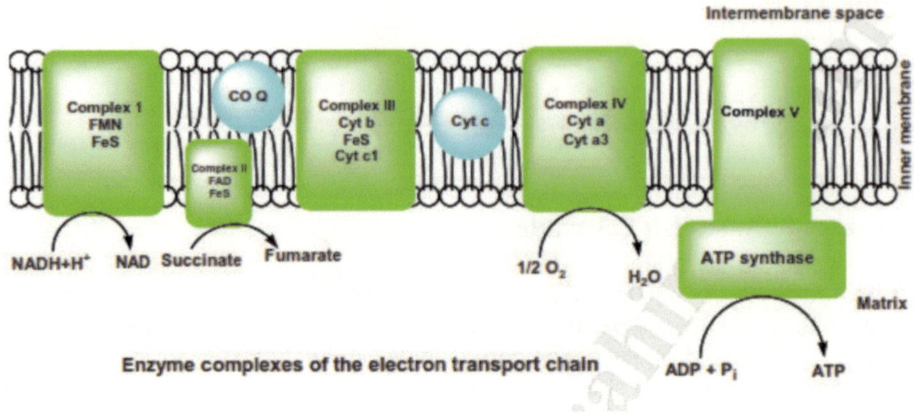

Figure 1.
Oxidoreductase enzymes involved in electron transport chain and oxidative phosphorylation [18].

6. Oxidoreductase in drug metabolism

Liver is the principal organ for drug metabolism. The body uses different strategies to metabolize drugs like oxidation, reduction, hydrolysis, hydration, conjugation, condensation, or isomerization. The main goal of drug metabolism is to make the drug more hydrophilic and excrete easily. Enzymes involved in drug metabolism are found in many tissues and organs but are more concentrated in the liver. Rates of drug metabolism may vary among individuals. Some individuals metabolize a drug so rapidly; in others, metabolism may be so slow and have different effects. Genetic factors, coexisting disorders (particularly chronic liver disorders and advanced heart failure), and drug interactions are responsible factors for variation of rate of drug metabolism among individuals [26].

Generally, drug metabolism can be in three phases. In phase I drug metabolism, oxidoreductase enzymes such as cytochrome P450 oxidases add polar or reactive groups into drugs (xenobiotics). In phase I reaction, drugs are introduced into new or modified functional group through oxidation, reduction, and hydrolysis. In Phase II reactions, modified compounds are in conjugation with an endogenous substance, e.g., glucuronic acid, sulfate, and glycine. Phase II reactions are synthetic, and compounds become more polar and thus, more readily excreted by the kidneys (in urine) and the liver (in bile) than those formed in nonsynthetic reactions. At the end, in phase III reaction, the conjugated drugs (xenobiotics) may be further processed, before being recognized by efflux transporters and pumped out of cells. The metabolism of drug often converts hydrophobic compounds into hydrophilic products that are more readily excreted [27].

In normal cases, human body wants to remove or detoxify any compounds that cannot be metabolized otherwise utilized to serve the needs of the body. This removal process is carried out mainly by the liver. The liver has classes of oxidoreductase enzymes that are extremely effective at detoxification and removal of drugs from the body [5, 18].

6.1 Metabolism of drugs through cytochrome P450 monooxygenase

Oxidation and metabolism of a high number of drugs and endogenous molecules are catalyzed by a class of oxidoreductase enzymes called cytochrome P450 monooxygenases. Even though they are distributed throughout the body, cytochrome

P450 enzymes are primarily concentrated in liver cells. The CYP2D6 isozymes play a great role in metabolizing certain opioids, neuroleptics, antidepressants, and cardiac medications. Currently it is going to be understood that difference in the genes for CYP450 enzymes play to inter-individual differences in the serum concentrations of drug metabolites, resulting in interpatient variability in drug efficacy and safety [28].

6.2 Metabolism of drugs with flavin-containing monooxygenase (FMO) system

Flavin-containing monooxygenases (FMOs) (EC 1.14.13.8) are a family of microsomal NADPH-dependent oxidoreductase, responsible for oxygenation of nucleophilic nitrogen, sulfur, phosphorus, other drugs, and endogenous molecules. Different variants of mammalian FMOs play a significant role in the oxygenation of nucleophilic xenobiotics. FMO utilizes NADPH as a cofactor and contains one FAD as a prosthetic group. FMOs have a broad substrate specificity and their activity is maximal at or above pH 8.4. FMO is a highly abundant enzyme in the liver endoplasmic reticulum and participates in drug metabolism (activation and detoxification) [29].

Before FMOs bind to a substrate, they activate molecular oxygen. First, flavin adenine dinucleotide (FAD), the prosthetic group of FMO, is reduced by NADPH to form FADH, then oxygen is added into the FAD, and hydro-peroxide FADH-4α-OOH is produced. And then, one oxygen atom is transferred to the substrate [30, 31].

6.3 Metabolism of drugs through alcohol dehydrogenase and aldehyde dehydrogenase

Alcohol dehydrogenase (ADH) and mitochondrial aldehyde dehydrogenase (ALDH) are another family of oxidoreductase responsible for metabolizing ethanol. These enzymes are highly expressed in the liver but at lower levels in many tissues and play a great role in detoxification and easy removal of alcohols. Liver is the main organ for ethanol metabolism. Oxidation of ethanol with these enzymes can become a major energy source especially in the liver, and it can interfere metabolism of other nutrients [32].

The first step in ethanol metabolism is its oxidation to acetaldehyde, and this reaction is catalyzed by enzymes called alcohol dehydrogenases (ADHs). The second reaction in ethanol metabolism is oxidation of acetaldehyde into acetate catalyzed by aldehyde dehydrogenase (ALDH) enzymes. There are different ADH and ALDH enzymes encoded by different genes occurring in several alleles and enzymes that have different alcohol metabolizing capacity; thereby, they influence individuals' alcoholism risk. These are either through rapid oxidation of ethanol to acetaldehyde where there is more active ADH or slower oxidation of acetaldehyde into acetate where there are less active ALDH enzymes. Excess accumulation of acetaldehyde is toxic, which results in different adverse reactions and produces nausea, skin rash, rapid heartbeat, etc. Most commonly, single-nucleotide polymorphisms (SNPs) are responsible for ADH and ALDH gene variants, and these may occur on both coding and non-coding regions of the gene [33, 34].

6.4 Metabolism of drugs by monoamine oxidase (MAO)

Monoamine oxidase is a very important oxidoreductase enzyme mainly responsible for degradation of amine neurotransmitters like norepinephrine, epinephrine, serotonin, and dopamine. Oxidation of different endogenous and exogenous biogenic amines may produce other active or inactive metabolites. Monoamine oxidase (MAO) is found in two isozyme forms: monoamine oxidase A (MAO-A) preferentially deaminates serotonin, norepinephrine, epinephrine, and dietary vasopressors such as tyramine, and MAO-B preferentially deaminates dopamine and phenethylamine. They are integral flavoproteins components of outer mitochondrial membranes in neurons and glia cell. The two isozymes of MAO differ based on substrate specificity and sensitivity to different inhibitors [35].

Monoamine oxidase enzymes catalyze the primary catabolic pathway for 5-HT oxidative deamination. Serotonin is converted into 5-hydroxy-indoleacetaldehyde, and this product is further oxidized by a NAD-dependent aldehyde dehydrogenase to form 5-hydroxyindoleacetic acid (5-HIAA). Immunohistochemical techniques and in situ hybridization histochemistry techniques are used to study the neuroanatomical localization and biochemical nature of the two forms of MAO [36].

Different antidepressant drugs like phenelzine and tranylcypromine inhibit the activity of monoamine oxidase. These are a result of MAO metabolizes biogenic amines such as 5-HT, DA, and NE. In addition, different dopaminergic neurotoxins such as 1-methyl-4-phenyl-1,2,3,6-tetrahydropyridine (MPTP) are metabolized by MAO [37].

6.5 NADPH-cytochrome P450 reductase (CPR) in drug metabolism

Another essential class of oxidoreductase enzyme is NADPH-cytochrome P450 reductase (CPR). It is a membrane-bound protein localized in the ER membrane. PR involves in the detoxification and activation of a number of xenobiotics. CPR uses FAD and FMN as cofactors, and it transfers the hydride ion of NADPH to FAD, and then FAD transfers electrons to FMN and other oxidases. Finally, it reduces the P450 enzyme heme center to activate molecular oxygen. Thus, electrons transfer from NADPH to the P450 heme center by CPR, which is central for P450-catalyzed metabolism. Flow of electron can be expressed as follows:

$$\text{NADPH} \rightarrow \text{FAD} \rightarrow \text{FMN} \rightarrow \text{P450} \rightarrow O_2 \tag{7}$$

Human cytochrome P450 reductase is encoded by the POR gene. It is a 78-kDa multi domain diflavin reductase that binds both FMN and FAD and is attached to the cytoplasmic side of the endoplasmic reticulum via a transmembrane segment at its N-terminus [5, 15, 38].

7. Industrial application of oxidoreductase enzymes

Several industries such as pharmaceutical, foods, biofuel production, natural gas conversion, and others have used enzyme catalysis at commercial scale [39]. Classes of oxidoreductase enzymes are becoming a target by a number of industries. The family of oxidoreductase like heme-containing peroxidases and peroxygenases, flavin-containing oxidases and dehydrogenases, and different copper-containing oxidoreductases is involved in synthesis and degradation of interested products by the above industries and they are biocatalysts of interest for establishing a bio-based economy. Oxidoreductase enzymes have the highest potential in the production of polymer building blocks, sustainable chemicals, and materials from plant biomass within lignocellulose biorefineries [6, 7, 40].

7.1 Oxidoreductase enzymes in pharmaceutical industries

Enzymes are biological catalysts and have great specificity, efficiency, and selectivity in the reaction they catalyze [39]. Oxidoreductase enzymes have different redox-active centers for doing their functions. These unique features of oxidoreductase enzymes make it valuable targets of pharmaceutical and chemical industries. Advancement in recombinant DNA technology, protein engineering, and bioinformatics is a critical event in the application of enzymes in different industries. A number of dug synthesis processes require the involvement of oxidoreductase enzymes [6].

An oxidoreductase is involved in the synthesis of 3,4-dihydroxylphenyl alanine (DOPA), and 3,4-dihydroxylphenyl alanine is a drug used for treatment of Parkinson's disease [41]. Similarly, a class of oxidoreductase called monoamine oxidase (MAO) catalyzes enantiomeric desymmetrization of bicyclic proline intermediate, which is an important precursor in the synthesis of boceprevir. Boceprevir is a NS3 protease inhibitor that is used for the treatment of chronic hepatitis C infections. Using MAO in this reaction reduces time and waste product generation and is economically cost-competitive and profitable [42]. Its coenzyme specificity makes oxidoreductase an effective biocatalyst in protein engineering [43]. In vitro different oxidoreductase enzymes are involved in regeneration of coenzymes, pyridine nucleotides, NAD(H) and NADP(H). Alcohol dehydrogenase and format dehydrogenase are frequently used enzymes for recycling of coenzymes, and the intermediate products are useful in the synthesis of pharmaceutical drugs such as mevinic acid [44, 45].

7.2 Oxidoreductase enzymes in agricultural sector

Enzymes are biological catalysts and have a number of applications in agricultural fields. Using enzymes has great efficacy and efficiency over chemical catalysts with respect to their productivity, time, cost, quality, and quantity products. There are different classes of oxidoreductase enzymes nowadays involved in fertilizer production, dairy processing, and other food processing in agricultural sector, and their cost-effectiveness and quality product were confirmed by a number of researches [3].

Manipulation of gene cod for different oxidoreductase in plants can also change the characters of plants in a way that it increases productivity and resists adverse effects of herbicide and environmental changes. For example, modification of DNA for glyphosate oxidoreductase (GOX) enzyme that catalyzes the oxidative cleavage of the C—N bond on the carboxyl side of glyphosate, resulting in the formation of aminomethylphosphonic acid (AMPA) and glyoxylate thereby augmented expression of GOX plants, results in glyphosate herbicide side effect tolerance [46, 47]. Some families of oxidoreductase like xanthine dehydrogenase in plants are used to metabolize reactive oxygen species associated with plant-pathogen and protect plants from stress-induced oxidative damage. Upregulation of xanthine dehydrogenase expression in plants is helpful to increase productivity [48, 49].

Classes of oxidoreductase are also involved in dairy processing. Glucose oxidase produced by fungal species acts as preservatives in dairy products and other foods. The intermediate and end product of glucose oxidase have antimicrobial effect [50]. Isozyme of xanthine oxidoreductase in bovine milk, which catalyzes reduction of oxygen to generate reactive metabolite is used as an anti-microbial agent in the neonatal gastrointestinal tract [51]. Similarly, peroxidases which are a family of oxidoreductase found in higher plants catalyze the oxidation of many compounds including phenolics, in the presence of hydrogen peroxide responsible in browning or darkening of noodles and pasta and associated with a grain quality defect [52]. Protochlorophyllide oxidoreductase (POR), which exists in two isozymes POR A and POR B, plays a vital role in plant chlorophyll synthesis, and manipulation on these genes can induce plant development [53]. In general, there are a number of oxidoreductase enzymes found in plants, and their normal activity is crucial for qualitative and quantitative productivity of crops, and these were confirmed by a number of active researches. Different interventions are also going on at gene level to control the expression of oxidoreductase enzymes in plant as needed [3].

8. Disease related with oxidoreductase enzyme disorder

Oxidoreductase enzymes are involved in a number of valuable biochemical reactions in the living organism, and their qualitative and quantitative normality is essential. For example, one important class of oxidoreductase is xanthine oxidoreductase (XOR) that catalyzes oxidative hydroxylation of hypoxanthine to xanthine then to uric acid and over activity XOR leads to hyperuricemia and concomitant production of reactive oxygen species. In turn, hyperuricemia is confirmed as an independent risk factor for a number of clinical conditions such as gout, cardiovascular disease, hypertension, and others. Different urate-lowering drugs or XOR inhibitors are nowadays implemented to prevent and manage hyperuricemia disorder [9].

Another important class of oxidoreductase enzyme is cytochrome P450 oxidoreductase (POR) that is essential for multiple metabolic processes. Cytochrome P450 enzymes are involved in metabolism of steroid hormones, drugs, and xenobiotics. Nowadays, more than 200 different mutations and polymorphisms in POR gene have been identified and cause a complex set of disorders. Deficiency of cytochrome P450 oxidoreductase affects normal production of hormone; specifically, it affects steroid hormones, which are needed for normal development and reproduction. This is highly linked with the reproductive system, skeletal system, and other functions. Signs and symptoms can be seen from birth to adult age with different severities. Individuals with moderate cytochrome P450 oxidoreductase deficiency may have ambiguous external genitalia and have a high chance of infertility but a normal skeletal structure [5, 16, 18].

Aldehyde dehydrogenase 2 (ALDH2) deficiency known as Asian glow or alcohol flushing syndrome is a common genetic health problem that interferes with alcohol metabolism, and ALDH2 is a classical family of oxidoreductase enzymes. It was confirmed that ALDH2 deficiency results in the accumulation acetaldehyde, which is a toxic metabolite of alcohol metabolism and responsible for a number of health challenges like esophageal, head, and neck cancer. A number of researches conclude that acetaldehyde is a group 1 carcinogenic metabolite [33, 54]. Similarly, monoamine oxidase deficiency, which is a family oxidoreductase enzyme, affects the normal metabolism of serotonin and catecholamines. It is a rare X-linked disorder characterized by mild intellectual disability, and behavioral challenges appear at earlier age. Monoamine oxidase-A deficiency that occurs almost exclusively in males has episodes of skin flushing, excessive sweating, headaches, and diarrhea. Monoamine oxidase-A deficiency can be diagnosed by finding an elevated urinary concentration of the monoamine oxidase-A substrates in combination with reduced amounts of the monoamine oxidase products [36, 55].

Mitochondria generate huge amounts of energy (ATP) to eukaryotic cells through oxidation of fats and sugars; and fatty acid β-oxidation and oxidative phosphorylation are two metabolic pathways that are central to this process. Qualitative and quantitative normality of oxidoreductase enzymes involved in oxidative phosphorylation and fatty acid oxidations are essential to get sufficient energy (ATP) form metabolism. Deficiency of a complex I (NADH-CoQ oxidoreductase) is common, and a well-characterized mitochondrial problem causes reduced ATP production [56]. Complex I (NADH-CoQ oxidoreductase) is responsible for recycling of NADH to NAD^+, and in turn, this is essential to sustain Krebs cycle and glycolysis. Mutations in both nuclear and mitochondrial DNA for Complex I gene are responsible for mitochondrial disease. Individuals with mitochondrial diseases suffer from an energy insufficiency characterized by myopathies, neuropathy, delayed development, cardiomyopathy, lactic acidosis, and others. Furthermore, since mitochondria are a hub of metabolism, mitochondrial dysfunctions are highly associated with metabolic diseases like hypertension, obesity, diabetes, neurodegenerative diseases, and even aging. Deficiency of complex I leads to elevation of NADH levels in the mitochondria that inhibit pyruvate dehydrogenase and α-ketoglutarate dehydrogenase. This condition completely inhibits Krebs cycle, and it is measured by CO_2 evolution from $[^{14}C]$ labeled precursors. Similarly, complex II (succinate:ubiquinone oxidoreductase) deficiency affects both fatty acid oxidation and electron transport chain, and it induces retinopathies and encephalopathies [57, 58].

Deficiency of the pyruvate dehydrogenase complex (PDHC), another class of oxidoreductase enzymes, causes similar clinical and biochemical alteration in energy production with complex I (NADH-CoQ oxidoreductase) [59]. Both TCA cycle and respiratory chain can be affected by succinate dehydrogenase deficiency. Deficiency of oxidoreductase enzymes involved in Krebs cycle affects all carbohydrate, protein, fat, and nucleic acid metabolism as it is a common pathway for metabolism of the above macromolecules [60].

Oxidoreductase enzymes are also involved in bile acid synthesis. Classes of oxidoreductase enzymes called 3beta-hydroxy-Delta (5)-C (27)-steroid oxidoreductase catalyze an early step of bile acids synthesis from cholesterol and are encoded by HSD3B7 gene on chromosome 16p11.2-12. Mutations of HSD3B7 gene affect bile acids synthesis, cause development of progressive liver disease characterized by cholestatic jaundice, malabsorption of lipids, and lipid-soluble vitamins from the gastrointestinal tract, and finally progress to cirrhosis and liver failure [61].

One important biomolecule that acts as a precursor for other molecules and a component of cell membrane is cholesterol. Mammalian cells can get cholesterol from de novo biosynthesis or uptake of exogenously derived cholesterol associated with plasma low-density lipoprotein (LDL). 3-Hydroxy-3-methylglutaryl-coenzyme A (HMG-CoA) reductase, which is a class of oxidoreductase, catalyzes the rate-limiting steps of de novo cholesterol biosynthetic pathway and target for manipulation pharmacologically. Under or over activity of HMG-CoA reductase can disturb cholesterol homeostasis and lead to either hypercholesterolemia or hypocholesterolemia. And disturbed cholesterol level associated with number serious clinical problem like atherosclerosis [62, 63].

Conflict of interests

The authors declare that they have no competing interests.

Authors' contributions

Mezgeu Legesse Habte drafted the paper and write the literature review.

Etsegenet Assefa assisted in guidance, critical assessment and peer review of the writing. Both authors have given their final approval of this version to be published. Both authors read and approved the final manuscript.

Ethical statement

Availability of data and material: All necessary data and materials related to the article are included in the article.

Funding: This review article is not funded by any person or organization (not funded).

Author details

Mezgebu Legesse Habte[1*] and Etsegenet Assefa Beyene[2]

1 Department of Biochemistry, Harmaya University, School of Medicine and College of Health Sciences, Ethiopia

2 Department of Biochemistry, College of Health and Medical Sciences, Addis Ababa University, Addis Ababa, Ethiopia

*Address all correspondence to: mezgebulegesse@gmail.com

References

[1] Trisolini L, Gambacorta N, Gorgoglion R, et al. FAD/NADH dependent oxidoreductases: From different amino acid sequences to similar protein shapes for playing an ancient function. Journal of Clinical Medicine. 2019;**8**:2117

[2] McDonald A. The Enzyme List Class 1—Oxidoreductases. Nomenclature Committee of the International Union of Biochemistry and Molecular Biology (NC-IUBMB); 2019

[3] Gramss G, Rudeschko O. Activities of oxidoreductase enzymes in tissue extracts and sterile root exudates of three crop plants, and some properties of the peroxidase component. New Phytology. 1998;**138**:401-409

[4] Braune A, Gütschow M, Blau M. An NADH-dependent reductase from *Eubacterium ramulus* catalyzes the stereospecific heteroring cleavage of flavanones and flavanonols. Applied and Environmental Microbiology. 2019. DOI: 10.1128/AEM.01233-19

[5] Rendic S, Guengerich FP. Survey of human oxidoreductases and cytochrome P450 enzymes involved in the metabolism of xenobiotic and natural chemicals. Chemical Research in Toxicology. 2015;**28**:38-42

[6] Martíneza AT, Ruiz-Dueñasa FJ, Camareroa S, et al. Oxidoreductases on their way to industrial biotransformations. Biotechnology Advances. 2017;**35**:815-831

[7] Chapman J, Ismail AE, Dinu CZ. Industrial applications of enzymes: Recent advances, techniques, and outlooks. Catalysts. 2018;**8**:238

[8] Berry CE, Hare JM. Xanthine oxidoreductase and cardiovascular disease: Molecular mechanisms and pathophysiological implications. The Journal of Physiology. 2004;**555**(3): 589-606

[9] Chen C, Lü J-M, Yao Q. Hyperuricemia and XOR inhibitors. Medical Science Monitor. 2016;**22**: 2501-2512

[10] Raja MMM, Raja A, Imran MM, Santha AMI, Devasena K. Enzyme application in diagnostic application. Biotechnology. 2011;**10**(1):51-59

[11] Hoidal JR, Xu R, Huecksteadt T, Sanders KA, Pfeffer K, Sturrock AB. Lung injury and oxidoreductases. Environmental Health Perspectives. 1998;**106**(5):1235-1239

[12] Uravleva GF. Histochemical investigation of the activity of oxidoreductases in the skin lesions of lepromatous leprosy patients. Bulletin of the World Health Organization. 1972; **46**:813-819

[13] Pey AL, Megarity CF, Timson DJ. NAD(P)H quinone oxidoreductase (NQO1): An enzyme which needs just enough mobility, in just the right places. Bioscience Reports. 2019;**39**(1)

[14] Brass EP, Hiatt WR, Gardner AW, Hoppe CL. Decreased NADH dehydrogenase and ubiquinolcytochrome c oxidoreductase in peripheral arterial disease. American Journal of Physiology. Heart and Circulatory Physiology. 2001; **280**:H603-H609

[15] Fang Y, Gao N, Tian X, Zhou J, et al. Effect of P450 oxidoreductase polymorphisms on the metabolic activities of ten cytochrome P450s varied by polymorphic CYP genotypes in human liver microsomes. Cellular Physiology and Biochemistry. 2018;**47**: 1604-1616

[16] Flück CE, Tajima T, Pandey AV, et al. Mutant P450 oxidoreductase

causes disordered steroidogenesis with and without Antley-Bixler syndrome. Nature Genetics. 2004;**36**(3)

[17] Younus H. Oxidoreductase: An overview and practical application. In: Biocatalysis. 2019. pp. 39-55

[18] Zanger UM, Schwab M. Cytochrome P450 enzymes in drug metabolism: Regulation of gene expression, enzyme activities, and impact of genetic variation. Pharmacology & Therapeutics. 2013; **138**:103-141

[19] Ramsay RR. Electron carriers and energy conservation in mitochondrial respiration. ChemTexts. 2019;**5**:9

[20] Porpaczy Z, Sumegi SB, Alkonyi I. Interaction between NAD-dependent isocitrate dehydrogenase, a-ketoglutarate dehydrogenase complex, and NADH:Ubiquinone oxidoreductase. Journal of Biochemistry. 1987;**262**(20): 950-9514

[21] Shestov AA, Liu X, Ser Z, et al. Quantitative determinants of aerobic glycolysis identify flux through the enzyme GAPDH as a limiting step. Biochemistry Human Biology and Medicine. 2014;**3**:e03342

[22] Mailloux RJ. Still at the center of it all; novel functions of the oxidative Krebs cycle. Bioenergetics. 2015;**4**:122. DOI: 10.4172/21677662.1000122

[23] Ponizovskiy MR. Role of Krebs cycle in the mechanism of stability internal medium and internal energy in an organism in norm and in mechanisms of cancer pathology. Modern Chemistry and Applications. 2016;**4**:191. DOI: 10.4172/2329-6798.1000191

[24] Kotb MI. Biological Oxidation (Electron transport chain—Chemiosmosis Oxidative Phosphorylation—Uncouplers). Presentation. Bio II-Spring; 2017

[25] Matsubayashi M, Inaoka DK, Komatsuya K, et al. Novel characteristics of mitochondrial electron transport chain from *Eimeria tenella*. Genes. 2019;**10**:29

[26] Ozougwu J. Physiology of liver. Research Gate. 2017;**4**:13-24

[27] Penner N, Woodward C, Prakash C. Drug metabolizing enzymes and biotransformation reactions. Oxidative Enzymes. 2012:546-565

[28] Schenkman JB. Historical Background and Description of the Cytochrome P450 Monooxygenase System. 1993

[29] Başaran R, Eke BC. Flavin containing monooxygenases and metabolism of xenobiotics. Turkish Journal of Pharmaceutical Sciences. 2017;**14**(1):90-94

[30] Basaran R, Benay CA. Flavin containing monooxygenases and metabolism of xenobioics. Turkish journal of pharmaceutical sciences. 2017;**14**(1):90

[31] Rossner R, Kaeberlein M, Leiser S. Flavin containing monooxygenases in aging and disease: Emerging roles for ancient enzymes. The Journal of Biological Chemistry. 2017;**292**(27): 11138-11146

[32] Haseba T, Ohno YA. New view of alcohol metabolism and alcoholism—Role of the high-Km Class III alcohol dehydrogenase (ADH3). International Journal of Environmental Research and Public Health. 2010;7(3):1076-1092

[33] Chang JS, Hsiao J-R, Chen C-H. ALDH2 polymorphism and alcohol-related cancers in Asians: A public health perspective. Journal of Biomedical Science. 2017;**24**:19

[34] Agarwal DP et al. Alcohol Metabolism, Alcohol Intolerance, and

Alcoholism. Berlin Heidelberg: Springer-Verlag; 1990

[35] Yeung AK, Georgieva MG, Atanasov AG, Tzvetkovetics NT. Monoamine oxidases (MAOs) as privileged molecular targets in neuroscience. Research literature analysis. Frontiers in molecular neuroscience. 2019;**12**:143

[36] Müller T, Möhr J-D. Pharmacokinetics of monoamine oxidase B inhibitors in Parkinson's disease. Drug Metabolism & Toxicology. 2019;**15**(5):429-435

[37] Zhang Z, Hamada H, Gerk PM. Selectivity of dietary phenolics for inhibition of human monoamine oxidases A and B. BioMed Research International. 2019;**23**:2019

[38] Scott RR, Miller WL. Genetic and clinical features of P450 oxidoreductase deficiency. Hormonal Research. 2008; **69**:266-275

[39] Abdelraheem EMM, Busch H, Hanefeld U, Tonin F. Biocatalysis explained: From pharmaceutical to bulk chemical production. Reaction Chemistry and Engineering. 2019;**4**:1878-1894

[40] Fernandes P. Enzymes in food processing: Condensed overview on strategies for better biocatalysts. Enzyme Research. 2010;**2010**(1):1-20

[41] Stefano AD, Marinelli L, Eusepi P, et al. Synthesis and biological evaluation of novel selenyl and sulfur-l-dopa derivatives as potential anti-Parkinson's disease agents. Biomolecules. 2019;**9**:239

[42] Bhalerao DS, Kumar A, Arkala R, et al. Synthesis and process optimization of boceprevir: A protease inhibitor drug. Organic Process Research and Development. 2015;**19**(11):1559-1567

[43] Howe AYM, Venkatraman S. The discovery and development of boceprevir: A novel, first-generation inhibitor of the hepatitis C virus NS3/4A serine protease. Journal of Clinical and Translational Hepatology. 2013;**1**:22-32

[44] Daiju D, Takumi O, Yosifumi M, et al. Formate oxidase, an enzyme of the glucose-methanol-choline oxidoreductase family, has a His-Arg pair and 8-formyl-FAD at the catalytic site. Bioscience, Biotechnology, and Biochemistry. 2011;**75**(9):1662-1667

[45] Chánique AM, Parra LP. Protein engineering for nicotinamide coenzyme specificity in oxidoreductases: Attempts and challenges. Frontiers in Microbiology. 2018;**9**:194

[46] Pollegioni L, Schonbrunn E, Siehl D. Molecular basis of glyphosate resistance —Different approaches through protein engineering. The FEBS Journal. 2011; **278**(16):2753-2766

[47] Kumar K, Gambhir G, Dass A, et al. Genetically modified crops: Current status and future prospects. Planta. 2020;**251**:91

[48] Zdunek-Zastocka E, Lips HS. Is xanthine dehydrogenase involved in response of pea plants (*Pisum sativum* L.) to salinity or ammonium treatment? Acta Physiologiae Plantarum. 2003;**25**: 395-401

[49] Hofmann NR. Opposing functions for plant xanthine dehydrogenase in response to powdery mildew infection: Production and scavenging of reactive oxygen species. The Plant Cell. 2016; **28**(5):1001

[50] Buchert J, Autio K. Using crosslinking enzymes to improve textural and other properties of food. Novel Enzyme Technology for Food Applications. 2007

[51] Di Maio G, Pittia P, Mazzarino L, Maraschin M, Kuhnen S. Cow milk enriched with nanoencapsulated

phenolic extract of jaboticaba (plinia peruviana). Journal of Food Science and technology. 2019;**56**(3):1165-1173

[52] Pandey VP, Awasthi M, Singh S, Tiwari S, Dwivedi UN. Comprehensive review on function and application of plant peroxidases. Journal of Biochemistry & Analytical Biochemistry. 2017;**6**:308

[53] Garrone A, Archipowa N, Zipfel PF, Hermann G, Dietzek B. Plant Protochlorophyllide Oxidoreductases A and B – Catalytic Efficiency and Initial Reaction Steps. Journal of biological chemistry. 2015;**290**(47):28530-28539

[54] Ho T, Chang C, Wu J, Huang I, et al. Recombinant expression of aldehyde dehydrogenase 2 (ALDH2) in *Escherichia coli* nissle 1917 for oral delivery in ALDH2-deficient individuals. bioRxiv. 2019. preprint

[55] Shi Y, van Rhijn JR, Bormann M, Mossink B, et al. Brunner syndrome associated MAOA dysfunction in human induced dopaminergic neurons results in dysregulated NMDAR expression and increased network activity. bioRxiv. 2019. preprint

[56] Nsiah-Sefaa A, McKenzie M. Combined defects in oxidative phosphorylation and fatty acid β-oxidation in mitochondrial disease. Bioscience Reports. 2016;**36** art: e00313

[57] Chen X, Qi F, Dash RK, Beard DA. Kinetics and regulation of mammalian NADH-ubiquinone oxidoreductase (Complex I). Biophysical Journal. 2010; **99**:1426-1436

[58] Scheffler IE. Mitochondrial disease associated with complex I (NADH-CoQ oxidoreductase) deficiency. Journal of Inherited Metabolic Disease. 2015;**38**(3): 405-415

[59] Gupta N, Rutledge C. Pyruvate dehydrogenase complex deficiency: An unusual cause of recurrent lactic acidosis in a paediatric critical care unit. Journal of Critical Care Medicine. 2019; **5**(2):71-75

[60] Dean B, Chrisp GL, Quartararo M, et al. P450 oxidoreductase deficiency: A systematic review and meta-analysis of genotypes, phenotypes, and their relationships. The Journal of Clinical Endocrinology and Metabolism. 2020; **105**(3):e42-e52

[61] Cheng JB et al. Molecular genetics of 3beta-hydroxy-delta5-C27-steroid oxidoreductase deficiency in 16 patients with loss of bile acid synthesis and liver disease. The Journal of Clinical Endocrinology and Metabolism. 2003; **88**(4):1833-1841

[62] Lacher S, Bruttger J, Kalt B, et al. HMG-CoA reductase promotes protein prenylation and therefore is indispensible for T-cell survival. Cell Death & Disease. 2017;**8**:e2824

[63] Göbel A, Breining D, Rauner M, Hofbauer LC, Rachner TD. Induction of 3-hydroxy-3-methylglutaryl-CoA reductase mediates statin resistance in breast cancer cells. Cell Death & Disease. 2019;**10**(2):91

Chapter 2

Applications of Oxidoreductases

Sandhya Rani Gogoi

Abstract

Oxidoreductases comprise of a large group of enzymes catalyzing the transfer of electrons from an electron donor to an electron acceptor molecule, commonly taking nicotinamide adenine dinucleotide phosphate (NADP) or nicotinamide adenine dinucleotide (NAD) as cofactors. Research on the potential applications of oxidoreductases on the growth of oxidoreductase-based diagnostic tests and better biosensors, in the design of inventive systems for crucial coenzymes regeneration, and in the creation of oxidoreductase-based approaches for synthesis of polymers and oxyfunctionalized organic substrates have made great progress. This chapter focuses on biocatalytic applications of oxidoreductases, since many chemical and biochemical transformations involve oxidation/reduction processes, developing practical applications of oxidoreductases has long been a significant target in biotechnology. Oxidoreductases are appropriate catalysts owing to their biodegradability, specificity and efficiency and may be employed as improved biocatalysts to substitute the toxic/expensive chemicals, save on energy/resources consumption, generate novel functionalities, or reduce complicated impacts on environment.

Keywords: oxidoreductases, cofactors, biosensors, coenzymes regeneration, biocatalytic

1. Introduction

The various chemical transformations catalyzed by enzymes make these catalysts a key goal for utilization by the promising biotechnology industries. In the recent years, intense research in the field of enzyme technology has provided numerous approaches that facilitate the practical application of enzymes. This chapter emphasizes the application of oxidoreductases which catalyze the exchange of electrons amid the donor and acceptor molecules, in reactions involving electron transfer, proton/hydrogen extraction, hydride transfer, oxygen insertion, or other imperative steps. Oxidoreductases acquire advantage from the inclusion of different cofactors - for instance heme, flavin and metal ions - to catalyze redox reactions [1]. Majority of oxidoreductases are nicotinamide cofactor-dependent enzymes which have a high preference for nicotinamide adenine dinucleotide phosphate (NADP) or nicotinamide adenine dinucleotide (NAD) and they are further classified in six major classes which are oxidases, dehydrogenases, hydroxylases, oxygenases, peroxidases and reductases [2]. This chapter demonstrates the potential applications of oxidoreductases on the growth of oxidoreductase-based diagnostic tests and better biosensors, in the design of inventive systems for crucial coenzymes regeneration, and in the formation of oxidoreductase-based approaches for synthesis of polymers and oxyfunctionalized organic substrates.

2. Oxidoreductase-based diagnostic tests and as biosensors

The diagnosis and monitoring of a variety of diseases is extremely demanding nowadays for routine examination of clinical samples and other associated tests. The diagnostic enzymes are used for the detection/diagnosis or prognosis of disease conditions due to their substrate specificity and quantitated activity in the presence of other proteins, and are preferred in diagnosis, which can be used as a diagnostic tool for disease detection [3]. Depending on the verity of the disease, diseased state often leads to tissue damage. In such conditions, enzymes specific to diseased organs are released into blood circulation with augmented enzyme activity. The measurement of corresponding enzyme activities in blood/plasma, or any other body fluid, has been exploited in the diagnosis of diseased tissues/organs [3].

Jixu Wang et al. [4] investigated the expression and significance of glucose-6-phosphate dehydrogenase (G6PD) in human gastric cancer progression and prognosis. Apoptosis and necrosis are two major types of cell death in normal and disease pathologies. A key signature for necrotic cells is the permeabilization of the plasma membrane which can be quantified in tissue culture settings by measuring the release of the intracellular enzyme lactate dehydrogenase (LDH). It has been described that the measuring LDH release is a useful method for the detection of necrosis [5]. Two dehydrogenases, specifically, sorbitol dehydrogenase (SDH) and LDH, are used for cancer prognosis [3]. Reports suggested that in prostate cancer [6], and precancerous colorectal neoplasms [7], an abnormal serum concentration of SDH has been observed. Additionally, an enhanced level of SDH can be observed in acute liver damage and parenchymal hepaticdiseases [3]. It has been reported that LDH, marker of anaerobic metabolism, is associated with highly invasive and metastatic breast cancer and suggested that the association of activity of LDH in tumor tissue with mammographic characteristics could help in defining aggressive breast cancers [8]. The gene expression of LDH is studied in several human malignant tumors, collectively among colorectal cancer [9], lung cancer [10–12], breast cancer [13], oral cancer [14], prostate cancer [15], germ cell cancer [16], and pancreatic cancer [17]. In recent times, the prognostic value of the serum LDH level in cancer patients has been considered as a significant area of research. Additionally, LDH performs as a prognostic marker in patients with acute leukemia [18] and sickle cell disease [19].

A biosensor is an analytical tool that comprises a biological or biologically derived sensing matter with close proximity to the physico-chemical transducer [3]. The chief function of such a device is to produce a discrete or uninterrupted signal that is comparative to the concentration of the analyte [20]. Enzyme-based chemical biosensors are based on biological recognition and in order to function, the enzymes must be accessible to catalyze a specific biochemical reaction and be stable under the normal operating circumstances of the biosensor [21]. Generally the function of oxidoreductase biosensors is dependent on charge transport amid the enzyme and an electrode surface by means of coenzymes or redox mediators [22].

Over the years, various enzyme-based biosensors have been developed, however only a few of them are commercialized. The majority of the published work on enzymatic biosensors focuses on targeted blood glucose monitoring based on amperometric techniques [3]. The earliest glucose biosensor based on glucose dehydrogenase from Erwinia sp. and carbon paste was generated by Laurinavicius et al. [23] where the enzyme was incorporated in a polylysine-albumin gel, and the anchoring material was a paste of chemically adapted carbon powder, fumed silica, and binding material. A cellulose dehydrogenase based glucose biosensor from a mutant of *Corynascus thermophilus* has been developed, and a glassy carbon electrode (GCE) was acquired

Enzymes	Analyte	Test sample	Disease diagnosed	References
Glucose oxidase	Glucose	Blood plasma, blood serum, urine, and saliva	Diabetes, hypoglycemia	[26–29]
Oxalate oxidase	Oxalate	Blood serum and urine	Idiopathic urolithiasis and various intestinal diseases	[30]
Cholesterol oxidase	Cholesterol	Blood serum	Coronary heart disease, myocardial and cerebral infarction (stroke)	[31–34]
Lactate oxidase	Lactate	Blood plasma, blood serum, drug and biological samples	Hyper lactatemia, cardiac arrest, resuscitation, sepsis, reduced renal excretion, decreased extra hepatic metabolism, intestinal infarction and lacticacidosis	[35–40]

Table 1.
Oxidoreductase enzymatic biosensors as diagnostic tools.

by direct electrode position of gold nanoparticles (AuNPs). The biosensor was used for the detection of glucose in human saliva samples, with successful results in terms of both revival and association with glucose blood levels [24]. This proposes the development of noninvasive glucose monitoring devices. The details of different oxidoreductase enzymatic biosensors applied for clinical diagnosis are listed in **Table 1**. The first marketable biosensor (glucose biosensor) was commenced in 1975 which was derived from the electrochemical recognition of hydrogen peroxide, and the glucose oxidase was employed for the improvement of the biosensor [3]. Subsequently, Clemens et al. [25] established a novel amperometric glucose biosensor in a bedside artificial pancreas, and it was marked underneath the brand name "Biostator" by Miles (Elkhart, Indiana).

3. Oxidoreductases in coenzymes regeneration

The most of oxidoreductases for catabolism and anabolism significantly require two natural nicotinamide-based coenzymes (NAD and NADP), respectively. The most NAD(P)-dependent oxidoreductases choose one coenzyme as an electron acceptor or donor to the other depending on their diverse metabolic functions [41]. Generally coenzymes are involved in these oxidoreductase-catalyzed reactions to transport electron, hydride, hydrogen, oxygen, or other atoms or small molecules in diverse enzymatic pathways [42, 43]. The nicotinamide adenine dinucleotide (NAD)/nicotinamide adenine dinucleotide phosphate (NADP), ubiquinone (CoQ), and flavin mononucleotide (FMN)/flavin adenine dinucleotide (FAD) are the typical coenzymes. Nicotinamide-based coenzymes for the electron transport and storage in the form of hydride groups are the most noteworthy in view of the fact that 80% of characterized oxidoreductases necessitate NAD as a coenzyme, and 10% of them require NADP as a coenzyme [44].

Nicotinamide coenzymes based dehydrogenases are of emergent importance for the production of chiral compounds, either by reduction of a prochiral precursor or via oxidative resolution of their racemate [45]. Nevertheless, the oxidized and reduced nicotinamide cofactors regeneration is an extremely critical step as the employ of these cofactors in stoichiometric amounts is too expensive for function. There are very few enzymes which are appropriate for the regeneration of oxidized nicotinamide cofactors. Glutamate dehydrogenase can be utilized for the oxidation of NADH in addition to NADPH while l-lactate dehydrogenase is able to oxidize NADH only [45]. The reduction of NAD^+ is carried out by formate and FDH [45]. Glucose-6-phosphate dehydrogenase and glucose dehydrogenase are proficient to reduce both NAD^+ and $NADP^+$ [45]. It has been reported that ADH from horse liver reduces NAD^+ whereas ADHs from *Lactobacillus* strains catalyze the reduction of $NADP^+$ [45]. These enzymes can be applied by their inclusion in entire cell biotransformations by an $NAD(P)^+$-dependent major reaction to achieve *in situ* regeneration of the consumed cofactor [45]. And for the regeneration of the reduced cofactors NADH and NADPH numerous systems for instance engineered formate dehydrogenase [46, 47], phosphite dehydrogenase [48, 49], glucose dehydrogenase [50, 51] plus cosubstrate are well established and extensively used.

Johannes et al. [52] reported the engineering of a highly stable and active mutant phosphite dehydrogenase (12x-A176R PTDH) from *Pseudomonas stutzeri* and evaluation of its potential as an effective NADPH regeneration system in an enzyme membrane reactor. They have utilized two practically imperative enzymatic reactions including xylose reductase-catalyzed xylitol synthesis and alcohol dehydrogenase-catalyzed (R)-phenylethanol synthesis as models, and the mutant PTDH was compared to the commercially available $NADP^+$-specific *Pseudomonas sp*. 101 formate dehydrogenase (mut Pse-FDH) that is extensively employed for NADPH regeneration [52]. Soluble water-forming NAD(P)H oxidases comprise a promising $NAD(P)^+$ regeneration scheme since they only require oxygen as cosubstrate and produce water as only byproduct [53]. In addition, the thermodynamic equilibrium of O_2 reduction is a significant driving force for mostly energetically unfavorable biocatalytic oxidations [53]. Petschacher et al. [53] presented the generation of an NAD(P)H oxidase with high activity for both cofactors, NADH and NADPH. Applicability for cofactor regeneration is shown for coupling with alcohol dehydrogenase from *Sphyngobium yanoikuyae* for 2-heptanone production.

4. Oxidoreductase-based approaches for synthesis of polymers and various organic substrates

Enzyme catalyzed oxidation reactions have achieved growing concern in biocatalysis recently, reflected also by numerous outstanding reviews on this topic reported in the last years [54–56]. The group of oxidoreductases, to which all enzyme catalyzing oxidoreduction reactions, comprises numerous groups of biocatalysts such as dehydrogenases, monooxygenases, dioxygenases, oxidases, peroxidases, etc. [55]. Moreover, the enzymatic oxidative polymerizations have advantages of using nontoxic catalysts and mild reaction conditions, and the specific enzyme catalysis affords regio- and chemoselective polymerizations to construct functional materials [57]. It has been reported that peroxidases with the use of hydrogen peroxide as oxidant efficiently induce the oxidative coupling of phenols to phenolic polymers, the majority of which are scarcely attained by conventional chemical catalysts [57]. In addition, it has been published that laccase and peroxidase are helpful for production of cross-linked polymers such as artificial urushi and biopolymer hydrogel [57]. Kobayashi [58] established that the enzymatic polymerization as to be an efficient method of polymer synthesis. The polymerization uses hydrolases and oxidoreductases as catalysts and this new method of polymer synthesis afforded natural polysaccharides like cellulose, amylose, xylan, and chitin, and unnatural polysaccharides catalyzed by a glycosidase from well-designed monomers, varied functionalized polyesters catalyzed by lipase from a variety of monomers, and poly-aromatics materials catalyzed by an oxidoreductase and an enzyme model complex from phenols and anilines [58]. Furthermore, vinyl polymerization has been initiated by oxidoreductase [58].

Marjanovic et al. [59] reviewed the oxidative oligomerization and polymerization of various arylamines, e.g., aniline, substituted anilines, aminonaphthalene and its derivatives, catalyzed by oxidoreductases, such as laccases and peroxidases, in aqueous, organic, and mixed aqueous organic monophasic or biphasic media. Owing to the nontoxicity of oxidoreductases and their elevated catalytic effectiveness, as well as high selectivity of enzymatic oligomerizations/polymerizations under gentle conditions by means of primarily water as a solvent and often resulting in minimal byproduct formation enzymatic oligomerizations and polymerizations of arylamines are environmentally friendly and considerably contribute to a "green" chemistry of conducting and redox-active oligomers and polymers [59].

It has been also established that oxidative enzymes comprise privileged catalysts in organic synthesis [60]. Environmentally benign reaction conditions with high selectivity are the most fascinating characteristic exhibited by these biocatalysts in contrast to classical metal-based reagents. de Gonzalo et al. [60] reviewed the new perspectives and concepts derived from oxidative enzymatic processes, involving oxidative C-C bond forming reactions, atroposelective oxidations, oxidative dynamic processes, interconnected reactions, cyclic deracemizations, oxidative desymmetrizations and artificial oxidative enzymes. Oxidoreductases comprise an imperative group of biocatalysts as they facilitate not merely the broadly used stereoselective reduction of aldehydes and ketones but also the less well exploited oxidation of alcohols and amines [53]. In addition, oxidoreductases catalyzed oxidations are utilized for production of chiral alcohols and amines by deracemization [54, 60–62]. It has been reviewed thoroughly that the oxidoreductases enable chemists to perform highly selective and efficient transformations ranging from simple alcohol oxidations to stereoselective halogenations of non-activated C-H bonds [63]. Mifsud et al. [64] demonstrated for the first time that catalytic water oxidation mediated by robust TiO_2 semiconductors can be productively coupled to oxidoreductases achieving photobiocatalytic redox reactions.

One of the major applications of oxidoreductase is a pharmaceutical synthesis of 3,4-dihydroxylphenyl alanine (DOPA), which is employed in the treatment of Parkinson's disease and the industrial process that synthesizes DOPA make use of the oxidoreductase polyphenol oxidase [65]. It has been reported that the enantioselective reduction of C-4-substituted 3,5-dixocarboxylates can be carried out by using alcohol dehydrogenase from *Lactobacillus brevis* (LBADH) over-expressed in *E. coli* [66]. Laccase can be employed to synthesize numerous complex medicinal agents including triazolo(benzo)cycloalkyl thiadiazines, vinblastine, penicillin X dimer, cephalosporin antibiotics, and dimerized vindo-line [67]. In addition laccase can be used to synthesize a range of functional organic compounds including polymers with specific mechanical/electrical/optical properties, textile dyes, cosmetic pigments, flavor agents, and pesticides [68]. Biocatalysis is facilitating technology to organic synthesis chemistry by providing high selectivity of enzymatic reactions under mild conditions makes it a very valuable tool for green chemistry.

5. Medical applications

Due to the specificity and bio-based nature, potential applications of oxidoreductases in various fields are attracting active research efforts [69]. Several products generated by oxidoreductases are finding applications as antimicrobial, detoxifying, or active personal-care agents [69]. One potential application is laccase-based *in situ* generation of iodine, a reagent extensively used as disinfectant [67]. It has been described that laccase-iodide salt binary iodine-generating system (for sterilization) can have several advantages over the direct iodine application [69]. Peroxidases may replace laccase for the application, even though they would require H_2O_2 as cosubstrate [69]. The ClO^- and Mn(III) species formed by haloperoxidase and Mn-peroxidase are extremely effective oxidants and antimicrobial agents [70]. Peroxidase can also be used to cross-link collagen which is beneficial to the healing of damaged skin [71]. The physiological activities of lysyl oxidase comprise the extra-cellular matrix construction which can hasten wound-healing [72, 73]. A glucose oxidase, lactoperoxidase, and iodide system has been tested for dental care and the oxidase produces H_2O_2 to feed the peroxidase, so that it can produce iodine that can kill plaque-causing bacteria [74]. It has been reported that the haloperoxidase can be used to oxidatively modify rubber latex surfaces, making them less allergenic [75]. A secreted oxidoreductase may even be developed as a vaccine against secretor microbes such as, *Aspergillus oryzae* catalase A protein has been studied as a potential aspergillosis vaccine [69]. It has been reported that low-molecular-mass laccase purified from the mushroom *Tricholoma giganteumis* possesses significant HIV-1 reverse transcriptase inhibitory activity [76]. As nature's own catalysts, enzymes acquire very diverse specificity, reactivity, and other physicochemical, catalytic, and biological properties highly enviable for miscellaneous industrial and medical applications [69].

6. Conclusions

Tremendous progress has been made in the recent years in the field of applications of oxidoreductases. Oxidoreductases metabolism is a fundamental bio-process that plays a pivotal role in all species, including humans, plants, animals, and microorganisms, as their specific function is to catalyze oxidation and reduction reactions that occur within the cell. Abnormality in this metabolic system leads to a number of metabolic disorders. Thus, owing to the remarkable properties of oxidoreductases, they can be used for the diagnosis of disorders. They can provide insight into the diseased state by diagnosis, prognosis, or by assessment of response therapy. It has been established that oxidoreductases as biosensors are becoming popular potential tools in biotechnology due to their high specificity. With oxidoreductases, the conversion of a variety of aliphatic/aromatic molecules can be achieved; inert hydrocarbons can be functionalized (by hydroxylation, sulfoxidation, epoxidation, etc.); regio-, enantio- (on racemic substrates); enantiotopo– (on prochiral sub-strates); and chemo-selective reactions can be accomplished; important synthons from inexpensive and renewable biomaterials can be constructed; and the negative environment impact can be reduced [69]. Since numerous chemical and biochemical transformations engage oxidation/reduction processes, developing practical biocatalytic applications of oxidoreductases has long been an imperative target in biotechnology.

Acknowledgements

The author gratefully acknowledges the Department of Chemistry, Goalpara College (Assam), India.

Conflict of interest

The author declares no conflict of interest.

DOI: http://dx.doi.org/10.5772/intechopen.94409

Author details

Sandhya Rani Gogoi
Department of Chemistry, Goalpara College, Goalpara, Assam, India

*Address all correspondence to: gogoisandhyarani@gmail.com

References

[1] Martinez AT, Ruiz-Dueñas FJ, Camarero S, Serrano A, Linde D, Lund H, Vind J, Tovborg M, Herold-Majumdar OM, Hofrichter Mand Liers, C. Mint: Oxidoreductases on their way to industrial biotransformations. Biotechnology advances. 2017;35(6);815-831. DOI: https://doi.org/10.1016/j.biotechadv.2017.06.003

[2] Younus H. Oxidoreductases: Overview and Practical Applications. Biocatalysis: Springer, Cham; 2019. 39 p. DOI: https://doi.org/10.1007/978-3-030-25023-2_3

[3] Singh RS, Singh T, Singh AK. Enzymes as Diagnostic Tools. Advances in Enzyme Technology: Elsevier; 2019. 225p. DOI: https://doi.org/10.1016/B978-0-444-64114-4.00009-1

[4] Wang J, Yuan W, Chen Z, Wu S, Chen J, Ge J, Hou F, Chen Z. Mint: Overexpression of G6PD is associated with poor clinical outcome in gastric cancer. Tumor Biology. 2012;33; 95-101. DOI: https://doi.org/10.1007/s13277-011-0251-9

[5] Chan FKM, Moriwaki K, De-Rosa MJ. Detection of necrosis by release of lactate dehydrogenase activity. In: Snow A, Lenardo M. (Eds.), Immune Homeostasis Methods and Protocols. Springer Science +Bushiness Media, New York, 2013, vol. 979. p. 65-70. DOI: https://doi.org/10.1007/978-1-62703-290-2_7

[6] Szabo Z, Hamalainen J, Loikkanen I, Moilanen AM, Hirvikoski P, Vaisanen T, Paavonen TK, Vaarala MH. Mint: Sorbitol dehydrogenase expression is regulated by androgens in the human prostate. Oncology Reports. 2010;23;1233-1239. DOI: https://doi.org/10.3892/or_00000755

[7] Uzozie A, Nanni P, Staiano T, Grossmann J, Barkow-Oesterreicher S, Shay JW, Tiwari A, Buffoli F, Laczko E, Marra G. Mint: Sorbitol dehydrogenase over expression and other aspects of dysregulate dprotein expression in human precancerous colorectal neoplasms: a quantitative proteomics study. Molecular & Cellular Proteomics. 2014;13;1198-1218. DOI: https://doi.org/10.1074/mcp.M113.035105

[8] Radenkovic S, Milosevic Z, Konjevic G, Karadzic K, Rovcanin B, Buta M, Gopcevic K, Jurisic V. Mint: Lactate dehydrogenase, catalase and superoxide dismutase in tumor tissue of breast cancer patients in respect to mammographic findings. Cell Biochemistry and Biophysics. 2013;66;287-295. DOI: https://doi.org/10.1007/s12013-012-9482-7

[9] Koukourakis MI, Giatromanolaki A, Sivridis E, Gatter KC, Trarbach T, Folprecht G, Shi MM, Lebwohl D, Jalava T, Laurent D, Meinhardt G. Mint: Prognostic and predictive role of lactate dehydrogenase 5 expression in colorectal cancer patients treated with PTK787/ZK 222584 (vatalanib) antiangiogenic therapy. Clinical Cancer Research. 2011;17;4892-4900. DOI: 10.1158/1078-0432.CCR-10-2918

[10] Hermes A, Gatzemeier U, Waschki B, Reck M. Mint: Lactate dehydrogenase as prognostic factor in limited and extensive disease stage small cell lung cancer - a retrospective single institution analysis. Respiratory Medicine. 2010;104;1937-1942. DOI: https://doi.org/10.1016/j.rmed.2010.07.013

[11] Hsieh AH, Tahkar H, Koczwara B, Kichenadasse G, Beckmann K, Karapetis C, Sukumaran S. Mint: Pre-treatment serum lactate dehydrogenase as a biomarker in small cell lung cancer. Asia-Pacific Journal of Clinical Oncology. 2018;14(2);e64-70. DOI: https://doi.org/10.1111/ajco.12674

[12] Zheng X, Wang K, Xu L, Ye P, Cai S, Lu H, Bao C, Kong J. Mint: The effect of serum lactate dehydrogenase levels on lung cancer prognosis: ameta-analysis. International Journal of Clinical and Experimental Medicine. 2017;10;14179-14186. DOI: https://www.researchgate.net/publication/321028729

[13] Brown JE, Cook RJ, Lipton A, Coleman RE. Mint: Serum lactate dehydrogenase is prognostic for survival in patients with bone metastases from breast cancer: a retrospective analysis in bisphosphonate-treated patients. Clinical Cancer Research. 2012;18;6348-6355. DOI: 10.1158/1078-0432.CCR-12-1397

[14] Nandita A, Basavaraju SM, Pachipulusu B. Mint: Lactate dehydrogenase as a tumor marker in oral cancer and oral potentially malignant disorders: a biochemical study International Journal of Preventive & Clinical Dental Research. 2017;4;1-5. DOI: 10.5005/jp-journals-10052-0108

[15] Halabi S, Small EJ, Kantoff PW, Kattan MW, Kaplan EB, Dawson NA, Levine EG, Blumenstein BA, Vogelzang NJ. Mint: Prognostic model for predicting survival in men with hormone-refractory metastatic prostate cancer. Journal of Clinical Oncology. 2003;21;1232-1237. DOI: 10.1200/JCO.2003.06.100

[16] Gerlinger M, Wilson P, Powles T, Shamash J. Mint: Elevated LDH predicts poor outcome of recurrent germ cell tumours treated with dose dense chemotherapy. European Journal of Cancer. 2010;46;2913-2918. DOI: https://doi.org/10.1016/j.ejca.2010.07.004

[17] Rong Y, Wu W, Ni X, Kuang T, Jin D, Wang D, Lou W. Mint: Lactate dehydrogenase A is over expressed in pancreatic cancer and promotes the growth of pancreatic cancer cells. Tumor Biology. 2013;34;1523-1530. DOI: https://doi.org/10.1007/s13277-013-0679-1

[18] Walaa-Fikry ME. Mint: Lactate dehydrogenase (LDH) as prognostic marker in acute leukemia "Quantitative Method". Journal of Blood Disorders Transfusion. 2017;8;1-9. DOI: 10.4172/2155-9864.1000375

[19] Kato GJ, Nouraie SM, Gladwin MT. Mint: Lactate dehydrogenase and hemolysis in sickle cell disease, Blood. 2013;122;1091-1092. DOI: https://doi.org/10.1182/blood-2013-05-505016

[20] Turner APF, Karube I, Wilson GS. Biosensors: Fundamentals and Applications. 1st ed. Oxford University Press, Oxford, 1987. DOI: https://www.diva-portal.org/smash/get/diva2:619968/FULLTEXT01.pdf

[21] Rocchitta G, Spanu A, Babudieri S, Latte G, Madeddu G, Galleri G, Nuvoli S, Bagella P, Demartis MI, Fiore V, Manetti R. Mint: Enzyme biosensors for biomedical applications: Strategies for safeguarding analytical performances in biological fluids. Sensors. 2016;16(6);780-801. DOI: https://doi.org/10.3390/s16060780

[22] Schmidt HL, Schuhmann W. Mint: Reagentless oxidoreductase sensors. Biosensors and Bioelectronics, 1996;11(1-2);127-135. DOI: https://doi.org/10.1016/0956-5663(96)83720-1

[23] Laurinavicius V, Kurtinaitiene B, Liauksminas V, Ramanavicius A, Meskys R, Rudomanskis R, Skotheim T, Boguslavsky L. Mint: Oxygen insensitive glucose biosensor based on PQQ-dependent glucose dehydrogenase. Anal. Lett. 1999;32;299-316. DOI: https://doi.org/10.1080/00032719908542822

[24] Bollella P, Gorton L, Ludwig R, Antiochia R. Mint: A third generation glucose biosensor based on cellobiosedehydrogenase immobilized on a glassy carbon electrode decorated with electrodeposited gold nanoparticles:

characterization and application in human saliva. Sensors. 2017;17;1912-1926. DOI: https://doi.org/10.3390/s17081912

[25] Clemens AH, Chang PH, Myers RW. Mint: Development of an automatic system of insulin infusion controlled by blood sugar, its system for the determination of glucose and control algorithms. Journees annuelles de diabetologie de l'Hotel-Dieu. 1976; 269-278. DOI: https://pubmed.ncbi.nlm.nih.gov/1011418/

[26] Yao H, Li N, Xu JZ, Zhu JJ. Mint: A glucose biosensor based on immobilization of glucose oxidase in chitosan network matrix. Chinese Journal Chemistry. 2005;23;275-279. DOI: https://doi.org/10.1002/cjoc.200590275

[27] Chu X, Wu B, Xiao C, Zhang X, Chen J. Mint: A new amperometric glucose biosensor based on platinum nanoparticles/polymerized ionic liquid-carbon nanotubes nanocomposites. Electrochimica Acta. 2010;55;2848-2852. DOI: https://doi.org/10.1016/j.electacta.2009.12.057

[28] Periasamy AP, Chang YJ, Chen SM. Mint: Amperometric glucose sensor based on glucose oxidase immobilized on gelatin-multiwalled carbon nanotube modified glassy carbon electrode. Bioelectrochemistry. 2011;80;114-120. DOI: https://doi.org/10.1016/j.bioelechem.2010.06.009

[29] Qiu C, Wang X, Liu X, Hou S, Ma H. Mint: Direct electrochemistry of glucoseoxidase immobilized on nano structured gold thin films and its application to bioelectrochemical glucose sensor. Electrochimica Acta. 2012;67;140-146. DOI: https://doi.org/10.1016/j.electacta.2012.02.011

[30] Yadav S, Devi R, Kumari S, Yadav S, Pundir CS. Mint: An amperometric oxalate biosensor based on sorghum oxalate oxidase bound carboxylated multiwalled carbon nanotubes-polyaniline composite film. Journal of Biotechnology. 2011;151;212-217. DOI: https://doi.org/10.1016/j.jbiotec.2010.12.008

[31] Nandini S, Nalini S, Reddy MM, Suresh GS, Melo JS, Niranjana P, Sanetuntikul J, Shanmugam S. Mint: Synthesis of one-dimensional gold nanostructures and the electrochemical application of the nanohybrid containing functionalized graphene oxide for cholesterol biosensing. Bioelectrochemistry. 2016;110;79-90. DOI: https://doi.org/10.1016/j.bioelechem.2016.03.006

[32] Pakapongpan S, Tuantranont A, Sritongkham P. Mint: Cholesterol biosensor based on direct electron transfer of cholesterol oxidase on multi-wall carbon nanotubes. IEEE. 2011;2011;138-141. DOI: 10.1109/BMEiCon.2012.6172037

[33] Pundir CS, Narang J, Chauhan N, Sharma P, Sharma R. Mint: An amperometric cholesterol biosensor based on epoxy resin membrane bound cholesterol oxidase. Indian Journal of Medical Research. 2012;136;633-640. DOI: https://pubmed.ncbi.nlm.nih.gov/23168704/

[34] Sekretaryova AN, Beni V, Eriksson M, Karyakin AA, Turner AP, Vagin MY. Mint: Cholesterol self-powered biosensor. Analytical Chemistry. 2014;86;9540-9547. DOI: https://doi.org/10.1021/ac501699p

[35] Lupu A, Valsesia A, Bretagnol F, Colpo P, Rossi F. Mint: Development of a potentiometric biosensor based on nanostructured surface for lactate determination. Sensors Actuators B: Chemical. 2007;127;606-612. DOI: https://doi.org/10.1016/j.snb.2007.05.020

[36] Ibupoto ZH, Shah SMUA, Khun K, Willander M. Mint: Electrochemical

L-lactic acid sensor based on immobilized ZnO nanorods with lactate oxidase. Sensors. 2012;12;2456-2466. DOI: https://doi.org/10.3390/s120302456

[37] Haghighi B, Bozorgzadeh S. Mint: Fabrication of a highly sensitive electro chemiluminescence lactate biosensor using ZnO nanoparticles decorated multiwalled carbon nanotubes. Talanta. 2011;85;2189-2193. DOI: https://doi.org/10.1016/j.talanta.2011.07.071

[38] Jiang D, Chu Z, Peng J, Jin W. Mint: Screen-printed biosensor chips with Prussian blue nanocubes for the detection of physiological analytes. Sensors Actuators B: Chemical. 2016;228;679-687. DOI: https://doi.org/10.1016/j.snb.2016.01.076

[39] Romero MR, Garay F, Baruzzi AM. Mint: Design and optimization of a lactate amperometric biosensor based on lactate oxidase cross-linked with polymeric matrixes. Sensors Actuators B: Chemical. 2008;131;590-595. DOI: https://doi.org/10.1016/j.snb.2007.12.044

[40] Briones M, Casero E, Petit-Dominguez MD, Ruiz MA, Parra-Alfambra AM, Pariente F, Lorenzo E, Vazquez L. Mint: Diamond nanoparticles based biosensors for efficient glucose and lactate determination. Biosensors and Bioelectronics. 2015;68;521-528. DOI: https://doi.org/10.1016/j.bios.2015.01.044

[41] You C, Huang R, Wei X, Zhu Z, Zhang YHP. Mint: Protein engineering of oxidoreductases utilizing nicotinamide-based coenzymes, with applications in synthetic biology. Synthetic and Systems Biotechnology, 2017;2(3);208-218. DOI: https://doi.org/10.1016/j.synbio.2017.09.002

[42] Schomburg I, Jeske L, Ulbrich M, Placzek S, Chang A, Schomburg D. Mint: The BRENDA enzyme information system–From a database to an expert system. Journal of Biotechnology. 2017; 261;194-206. DOI: https://doi.org/10.1016/j.jbiotec.2017.04.020

[43] Wichmann R, Vasic-Racki D. Cofactor regeneration at the lab scale. In: Technology transfer in biotechnology. Springer, Berlin, Heidelberg; 2005. p. 225-260. DOI: https://doi.org/10.1007/b98911

[44] Wu H, Tian C, Song X, Liu C, Yang D, Jiang Z. Mint: Methods for the regeneration of nicotinamide coenzymes. Green Chemistry. 2013;15(7);1773-1789. DOI: 10.1039/C3GC37129H

[45] Weckbecker A, Gröger H, Hummel W. Regeneration of nicotinamide coenzymes: principles and applications for the synthesis of chiral compounds. In Biosystems Engineering I. Springer, Berlin, Heidelberg; 2010. p. 195-242. DOI: https://doi.org/10.1007/10_2009_55

[46] Tishkov VI, Popov VO. Mint: Protein engineering of formate dehydrogenase. Biomolecular Engineering. 2006;23;89-110. DOI: https://doi.org/10.1016/j.bioeng.2006.02.003

[47] Hoelsch K, Sührer I, Heusel M, Weuster-Botz D. Mint: Engineering of formate dehydrogenase: Synergistic effect of mutations affecting cofactor specificity and chemical stability. Applied Microbiology and Biotechnology. 2013;97;2473-2481. DOI: 10.1007/s00253-012-4142-9

[48] Johannes TW, Woodyer RD, Zhao H. Mint: Efficient regeneration of NADPH using an engineered phosphite dehydrogenase. Biotechnology and Bioengineering. 2007;96;18-26. DOI: https://doi.org/10.1002/bit.21168

[49] Woodyer R, Van der Donk WA, Zhao H. Mint: Relaxing the nicotinamide cofactor specificity of phosphite dehydrogenase by rational design. Biochemistry. 2003;42;11604-11614. DOI: https://doi.org/10.1021/bi035018b

[50] Wong C-H, Drueckhammer DG, Sweers HM. Mint: Enzymatic vs. fermentative synthesis: Thermostable glucose dehydrogenase catalyzed regeneration of NAD(P)H for use in enzymatic synthesis. Journal of the American Chemical Society. 1985;107;4028-4031. DOI: https://doi.org/10.1021/ja00299a044

[51] Kaswurm V, Hecke WV, Kulbe KD, Ludwig R. Mint: Guidelines for the application of NAD(P)H regenerating glucose dehydrogenase in synthetic processes. Advanced Synthesis and Catalysis. 2013;355;1709-1714. DOI: https://doi.org/10.1002/adsc.201200959

[52] Johannes TW, Woodyer RD, Zhao H. Mint: Efficient regeneration of NADPH using an engineered phosphite dehydrogenase. Biotechnology and Bioengineering. 2007;96(1);18-26. DOI: https://doi.org/10.1002/bit.21168

[53] Petschacher B, Staunig N, Müller M, Schürmann M, Mink D, De Wildeman S, Gruber K, Glieder A. Mint: Cofactor specificity engineering of Streptococcus mutans NADH oxidase 2 for NAD(P)$^+$ regeneration in biocatalytic oxidations. Computational and Structural Biotechnology Journal. 2014;9(14);e201402005. DOI: https://doi.org/10.5936/csbj.201402005

[54] Hollmann F, Arends IW, Buehler K, Schallmey A, Bühler B. Mint: Enzyme-mediated oxidations for the chemist. Green Chemistry. 2011;13(2);226-265. DOI: 10.1039/C0GC00595A

[55] Monti D, Ottolina G, Carrea G, Riva S. Mint: Redox reactions catalyzed by isolated enzymes. Chemical reviews, 2011;111(7);4111-4140. DOI: 10.1021/cr100334x

[56] Romano D, Villa R, Molinari F. Mint: Preparative biotransformations: oxidation of alcohols. ChemCatChem. 2012;4(6);739-749. DOI: https://doi.org/10.1002/cctc.201200042

[57] Uyama H. Synthesis of Poly(aromatic)s I: Oxidoreductase as Catalyst. In Enzymatic Polymerization towards Green Polymer Chemistry. Springer, Singapore. 2019. p. 267-305. DOI: https://doi.org/10.1007/978-981-13-3813-7_9

[58] Kobayashi S. Mint: Enzymatic polymerization: a new method of polymer synthesis. Journal of Polymer Science Part A: Polymer Chemistry. 1999; 37(16);3041-3056. DOI: https://doi.org/10.1002/(SICI)1099-0518(19990815)37:16<3041::AID-POLA1>3.0.CO;2-V

[59] Ćirić-Marjanović G, Milojević-Rakić M, Janošević-Ležaić A, Luginbühl S, Walde P. Mint: Enzymatic oligomerization and polymerization of arylamines: state of the art and perspectives. Chemical Papers. 2017;71(2);199-242. DOI: 10.1007/s11696-016-0094-3

[60] de Gonzalo G, A Orden, A, R Bisogno F. Mint: New trends in organic synthesis with oxidative enzymes. Current Organic Chemistry. 2012;16(21);2598-2612. DOI: https://doi.org/10.2174/138527212804004599

[61] Kroutil W, Mang H, Edegger K, Faber K. Mint: Biocatalytic oxidation of primary and secondary alcohols. Advanced Synthesis & Catalysis. 2004;346(2-3);125-142. DOI: https://doi.org/10.1002/adsc.200303177

[62] Turner NJ. Mint: Enantioselective oxidation of C–O and C–N bonds using oxidases. Chemical Reviews.

2011;111(7);4073-4087. DOI: https://doi.org/10.1021/cr200111v

[63] Dong J, Fernández-Fueyo E, Hollmann F, Paul CE, Pesic M, Schmidt S, Wang Y, Younes S, Zhang W. Mint: Biocatalytic oxidation reactions: A chemist's perspective. Angewandte Chemie International Edition. 2018; 57(30);9238-9261. DOI: https://doi.org/10.1002/anie.201800343

[64] Mifsud M, Gargiulo S, Iborra S, Arends IW, Hollmann F, Corma A. Mint: Photobiocatalytic chemistry of oxidoreductases using water as the electron donor. Nature Communications. 2014; 5(1);1-6. DOI: 10.1038/ncomms4145

[65] Paul PEV, Sangeetha V, Deepika RG. Emerging trends in the industrial production of chemical products by microorganisms. In: Recent developments in applied microbiology and biochemistry. Academic Press. 2019. p. 107-125. DOI: https://doi.org/10.1016/B978-0-12-816328-3.00009-X

[66] Wolberg M, Hummel W, Müller M. Mint: Biocatalytic reduction of β, δ-diketo esters: A highly stereoselective approach to all four stereoisomers of a chlorinated β, δ-dihydroxy hexanoate. Chemistry–A European Journal. 2001;5;7(21);4562-71. DOI: https://doi.org/10.1002/1521-3765(20011105)7:21<4562::AID-CHEM4562>3.0.CO;2-4

[67] Xu F. in The Encyclopedia of Bioprocessing Technology: Fermentation, Biocatalysis,and Bioseparation, eds. Flickinger MC, and Drew SW. 1545-1554 (John Wiley & Sons,New York, 1999).

[68] Ose T, Watanabe K, Mie T, Honma M, Watanabe H, Yao M, Oikawa H, Tanaka I. Mint: Insight into a natural Diels–Alder reaction from the structure of macrophomate synthase. Nature. 2003;422(6928);185-9. DOI: 10.1038/nature01454

[69] Xu F. Mint: Applications of oxidoreductases: recent progress. Industrial Biotechnology. 2005; 1;1(1);38-50. DOI: https://doi.org/10.1089/ind.2005.1.38

[70] Danielsen S, Christensen BE. PCT patent WO2003047351-A(2003).

[71] Miller DR, Tizard IR, Keeton JT, Prochaska JF. PCT patent WO200135882-A(2001).

[72] Ensley BD, inventor; Matrix Design, assignee. Wound healing compositions and methods using tropoelastin and lysyl oxidase. United States patent US 6,808,707. 2004 Oct 26. US patent US6808707-B2(2004).

[73] Perrier E, Cenizo V, Bouez C, Sommer P, Damour O, Gleyzal C, Andre V, Reymermier C, inventors; Centre National de la Recherche Scientifique CNRS, Coletica, assignee. Stimulation of the synthesis of the activity of an isoform of lysyl oxidase-like LOXL for stimulating the formation of elastic fibres. United States patent application US 10/852,065. 2004 Dec 16. US patent US2004253220-A1(2004).

[74] Szynol A, De Soet JJ, Sieben-van Tuyl E, Bos JW, Frenken LG. Mint: Bactericidal effects of a fusion protein of llama heavy-chain antibodies coupled to glucose oxidase on oral bacteria. Antimicrobial Agents and Chemotherapy. 2004;;48(9):3390-3395.

[75] Novozymes AS. Danish patent DK200100630-A (2001).

[76] Wang HX, Ng TB. Mint: Purification of a novel low-molecular-mass laccase with HIV-1 reverse transcriptase inhibitory activity from the mushroom Tricholoma giganteum. Biochemical and Biophysical Research Communications. 2004;315(2);450-4. DOI: 10.1016/j.bbrc.2004.01.064

Chapter 3

Role of Subcellular ROS in Providing Resilience to Vascular Endothelium

Sarah R. Aldosari, Maan A. Awad, Frank W. Sellke and Md. Ruhul Abid

Abstract

For decades, elevated levels of reactive oxygen species (ROS) have been associated with the pathogenesis of cardiovascular diseases (CVD), including myocardial ischemia and infarction (MI). However, several large clinical trials failed to demonstrate beneficial outcomes in response to the global reduction of ROS in patients with underlying CVD. Recent studies from our and other labs showed that it is rather a critical balance between mitochondrial and cytosolic ROS than total ROS levels which determines resilience of coronary endothelial cells (EC). Here, we will discuss published and unpublished work that has helped elucidate the molecular mechanisms by which subcellular ROS levels, duration and localization modulate metabolic pathways, including glycolysis and oxidative phosphorylation, energy production and utilization, and dNTP synthesis in EC. These redox-regulated processes play critical roles in providing resilience to EC which in turn help protect existing coronary vessels and induce coronary angiogenesis to improve post-MI recovery of cardiac function.

Keywords: endothelial cell metabolism, angiogenesis, vascular endothelial growth factor (VEGF), nitric oxide, reactive oxygen species (ROS), glycolysis, dNTP, fatty acid oxidation

1. Introduction

A single layer of endothelial cells (ECs) that covers the vascular lumen and plexus exhibits great plasticity to adapt to environmental cues [1, 2]. It is fascinating how the vascular system, the largest organ system of the body, connects all organs to secure adequate nutrients and blood supply. For that reason, maintaining vascular homeostasis is crucial for the health of the cardiovascular system. In a healthy body, although the ECs are an intricate, dynamic system, they appear to be in a quiescent state [1]. In pathological conditions such as ischemia and infarction, ECs rapidly switch phenotype to form new vessels in a process known as sprouting angiogenesis [3]. Reactive oxygen species (ROS) are believed to play crucial roles in determining the phenotype and fate of EC in both physiological and pathological conditions. Recent work has shown that a critical balance between mitochondrial and cytosolic ROS levels, but not global ROS levels, modulates endothelial function, EC metabolism and angiogenesis, and thus determines resilience of coronary EC [4–8].

IntechOpen

In this chapter, we will discuss the molecular mechanisms by which subcellular ROS modulate various metabolic pathways, regulate EC function and angiogenesis.

2. Reactive oxygen species in coronary endothelium

Previously, the pathology of cardiovascular diseases (CVD), including myocardial ischemia and infarction (MI), was believed to be associated with increased levels of ROS [4–8]. Recent studies show that it is rather a critical balance between cytosolic and mitochondrial ROS levels than total ROS levels which determine the resilience of coronary ECs in physiological as well as adverse conditions [6, 7, 9, 10].

ROS are produced in higher levels as a response to injuries by the cellular enzymes and mitochondria [8, 11]. ROS have been reported to contribute to the underlying pathology in almost all organs, and thus the notion that antioxidants would ameliorate pathological effects of ROS came into being. However, clinical trials failed to show beneficial effects of antioxidants in the treatment of CVD [12]. Other studies showed that decreased ROS levels had rather deleterious effects on CVD [6, 7]. Also, decreased ROS levels resulted in the inactivation of endothelial nitric oxide synthase (eNOS) and reduction in NO (nitric oxide) levels [11, 13]. Taken together, global reduction of ROS appeared to reduce endothelial resilience. It is crucial to note that ROS have paradoxical effects on ECs, and thus careful study of the levels, durations and sources of ROS while studying effects of ROS on EC will help advance our understanding of EC resilience during oxidative stress.

2.1 Source of reactive oxygen species in ECs

ROS are produced from different oxidoreductase enzymes and locations including NADPH oxidase, mitochondrial, xanthine oxidase, cytochrome P450 monooxygenase, and uncoupling of NOS [8, 11, 14]. In the vasculature, ECs rely on glycolytic pathways as their source of energy, thus NADPH oxidase enzymes appear to be the major source of ROS in both physiological and pathological conditions [14]. NADPH enzymes have different isoforms, and the major contributors are NOX1, NOX2, NOX4, and NOX5 [15, 16]. Recent studies showed importance of NOX2- and NOX4-derived ROS in endothelial survival or dysfunction, depending on their subcellular location and duration [8].

2.2 Endothelial NADPH oxidase as a major source of ROS in ECs

NADPH oxidase is an intracellular complex enzyme containing membrane-bound and cytosolic regulator subunits [14, 17, 18]. This enzyme produces ROS by transferring electrons from NAD(P)H to an oxygen molecule and is considered the major source of ROS in coronary endothelium. Distinct isoforms of NADPH enzymes have been shown to exhibit different physiological and pathological responses in vascular homeostasis.

NOX1 enzyme is primarily expressed in the vascular smooth muscle cells (VSMC) and it contributes to VSMC proliferation and migration [11, 19–21]. In disease conditions, NOX1 contributes to the impairment of endothelium-dependent vasorelaxation, as well as the augmentation of angiotensin II vasomotor response [11, 22, 23]. A study showed that NOX1-deficient mice attenuated the levels of ROS, neointimal growth, and migration. These findings suggest that the downregulation of NOX1 enzyme can prevent the formation of atherosclerotic plaque [15, 24]. Yet, further studies are warranted to explore the exact role of NOX1 in endothelial signaling.

In contrast, NOX2 enzyme has exhibited positive effects on coronary ECs. NOX2 enzyme stimulates the production of NO by the activation of AMPK-eNOS axis through Ca^{2+}-/calmodulin-dependent protein kinase kinase β (CaMKKβ) [6] resulting in coronary vasodilation, EC proliferation and migration. Although several studies support the beneficial effects of NOX2, they also exert detrimental effects on coronary EC depending on the duration of exposure. Short exposure of elevated ROS levels was associated with the previously mentioned pathway (i.e. CaMKKβ pathway). On the contrary, prolonged exposure of high ROS levels resulted in decrease bioavailability of NO, inactivation of mitochondrial antioxidant MnSOD [7, 8], and decreased EC proliferation and coronary vasodilatation.

NOX4 enzyme is abundant in human ECs [8] and produces H_2O_2 molecules rather than O_2^- [9, 11, 25–28]. NOX4 enzyme stimulates vascular angiogenesis through the activation of transforming growth factor β1 (TGF β1), and increases hemoglobin content [29]. NOX4-derived ROS cause vasodilation through endothelium hyperpolarization [30–32]. This occurs via the stimulation of endothelium Ca^{2+}-activated K^+ channel that causes the release of Ca^{2+} from the endoplasmic reticulum [29]. Additionally, NOX4 enzyme activates heme oxygenase-1 (HO-1), which confers a vascular protective response via different mechanisms [29]. Thus, therapeutic modalities that advocate for antioxidants in CVD needs careful consideration of the source and location of ROS.

Calcium-dependent NADPH oxidase, NOX5, is implicated in angiogenic response [33, 34]. It gets its name from its structure because it has an additional N-terminal region that binds to calcium [33]. This unique structure allows the enzyme activation through increased intracellular calcium. Similar to NOX4, NOX5 enzyme seems to produce predominantly H_2O_2 in ECs [24]. H_2O_2 has been implicated in the development of atherosclerotic plaque plausibly by increasing Ca^{2+} levels to promote eNOS-mediated NO synthesis and increasing nitroxide radicals [24]. One mechanism may include increased consumption of NO by ROS. Thus, it has been hypothesized that inhibition of NOX5 enzyme may show beneficial results by precluding oxidant injury to vascular EC.

NADPH enzyme isoforms have distinct locations and EC phenotypes. They have been shown to employ different physiological and pathological responses in vascular homeostasis. As discussed above, NOX1, NOX2, NOX4, and NOX5 are found in the vascular system and they contribute to endothelial resilience through several mechanisms. The roles of NADPH enzymes in physiological and pathological conditions have undergone a considerable evolution in recent years. However, further studies are necessary to deepen our understanding of their roles and contributions to EC resilience.

2.3 Endothelial mitochondrial ROS

Although oxidative phosphorylation in mitochondria play a major role in synthesizing energy in most tissues, EC primarily depends on anaerobic glycolysis for 85% of its ATP generation. ECs have fewer mitochondria and consume lower amounts of O2 than other cell types, and thus mitochondrial ROS are believed to be a minor source of ROS in EC in physiological conditions. However, recent studies demonstrated that sustained increase in NADPH oxidase-derived cytosolic ROS may affect the levels of mitochondrial ROS and thus mitochondrial function in EC [6–8].

3. Metabolic pathways in ECs

EC metabolism plays an important role in facilitating cellular proliferation and migration during the process of angiogenesis. Alterations in metabolic pathways are necessary to provide energy supplies in the most efficient way under certain circumstances that induces blood vessel sprouting such as hypoxia. In addition, these alterations mediate the formation of important molecules that are essential for cytoskeletal remodeling during the process. This section highlights some of these metabolic pathways and their role in angiogenesis.

3.1 Glycolysis

Glycolysis is a major metabolic pathway that is utilized for energy production through the anaerobic oxidation of glucose molecules [35, 36]. It is the major source of ATP in ECs. Glycolysis involves consumption of 2 ATP molecules, and the end products include 4 ATP, 2 NADH and 2 pyruvate molecules (**Figures 1** and **2**). Subsequently, pyruvate can be shifted to the mitochondria and metabolized into acetyl-CoA to be used in the tricarboxylic acid cycle (TCA). The substrates and products of this process are as follows:

$$\text{Glucose} + 2\,\text{NAD}^+ + 2\,\text{ADP} + 2\,\text{Pi} \rightarrow 2\,\text{Pyruvate} + 2\,\text{NADH} + 2\,\text{H}^+ + 2\,\text{ATP} + 2\,\text{H}_2\text{O}.$$

Glycolysis occurs in the cytosol, and the process does not require oxygen (anaerobic), therefore it constitutes the primary source of energy in cells that lack mitochondria (e.g. red blood cells). In addition, glycolysis is the main source of pyruvate, which is converted to acetyl-CoA to be utilized in the TCA cycle in cells

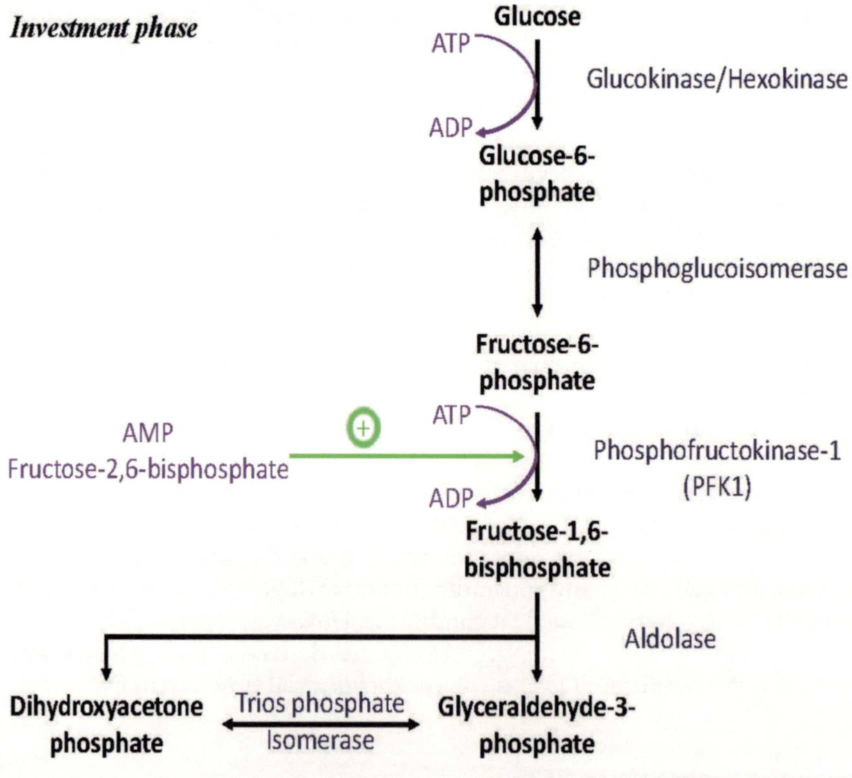

Figure 1.
The investment phase of glycolysis and regulation of the rate limiting PFK1 enzyme by fructose-2, 6-bisphosphate and AMP.

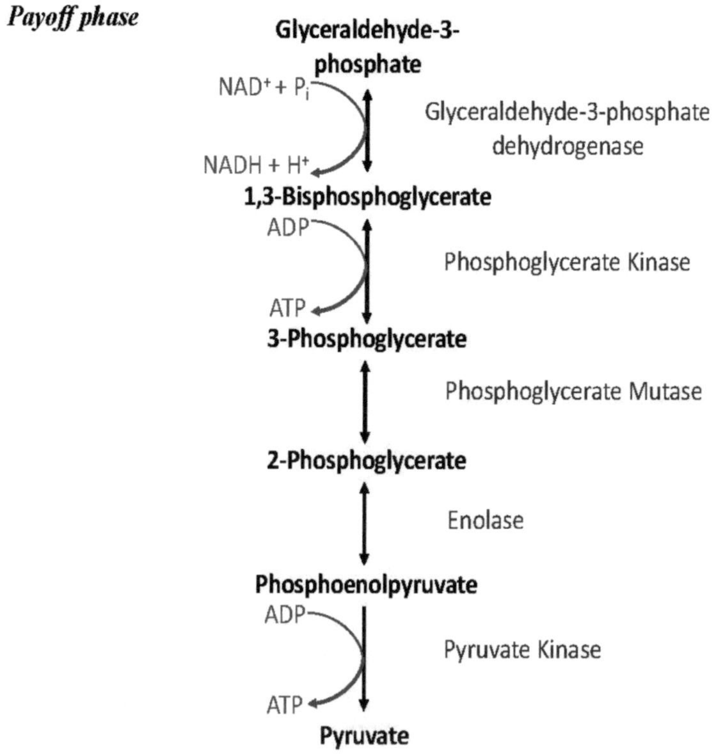

Figure 2.
The payoff phase of glycolysis.

that use oxidative phosphorylation (aerobic respiration) as a primary source of ATP. Also, glycolysis is a more efficient source of energy in periods of hypoxia and ischemia when oxygen supply becomes scarce.

3.1.1 Mechanisms of glycolysis

The first step in glycolysis constitutes the *investment* phase of glycolysis, in which 2 ATP molecules are consumed as shown in **Figure 1**. It involves trapping of the glucose molecule inside the cell via phosphorylation into glucose-6-phosphate [35, 36]. This reaction is catalyzed by glucokinase in the liver and pancreatic β cells, or a hexokinase enzyme in the rest of body cells. It also involves the transfer of a phosphate group from an ATP molecule. Next, glucose-6-phosphate is converted to fructose-6-phosphate by an isomerase. This is followed by the rate-limiting step of glycolysis, which involves the phosphorylation of fructose-6-phosphate into fructose-1-6-bisphosphate by phosphofructokinase 1 (PFK1). This step is critical in the glycolytic pathway and the PFK1 enzyme is highly regulated by multiple factors that determine the direction of the reaction. Fructose-1-6-bisphosphate is subsequently converted by an aldolase into dihydroxyacetone phosphate (DHAP) and glyceraldehyde-3-phosphate (G3P).

The following reactions constitute what can be referred to as the *payoff* phase of glycolysis. It is also important to remember that at this stage, we have two 3-carbon molecules per 1 glucose molecule as shown in **Figure 2**. G3P is converted to 1-3-diphosphoglycerate, generating NADH in the process. 1-3-diphosphoglycerate then loses a phosphate group to 3-phosphoglycerate via phosphoglycerate kinase, which generates an ATP molecule. 3-phosphoglycerate is subsequently converted in a two-step reaction into phosphoenolpyruvate (PEP). Finally, pyruvate kinase converts PEP into pyruvate, generating ATP in the process. Thus, the end product of glycolysis includes 4 ATP molecules, but because of the initial consumption of 2 ATP, the return on investment includes 2 ATP molecules per glucose [35, 36].

3.1.2 Regulation of glycolysis

The availability of glucose regulates the rate of glycolysis and is determined by two main mechanisms: glucose uptake from the blood, and breakdown of glycogen [35, 37]. In addition, the amount of oxygen can also regulate glycolysis through what is called the "Pasteur Effect", which describes how increased oxygen levels inhibit glycolysis, and decreased availability results in acceleration of glycolysis [35]. Within the glycolytic pathway, PFK1, which catalyzes the rate limiting step is considered the main player in terms of glycolysis regulation, and its activity can be affected in a number of ways.

Fructose 2–6 bisphosphate is an allosteric regulator of PFK1, which increases the enzyme activity [35, 37]. It is produced by phosphofructokinase 2 (PFK2), an enzyme that has both kinase and phosphorylase activity and can transform fructose 6 phosphates to fructose 2,6 bisphosphate and vice versa. Insulin dephosphorylates PFK2 activating its kinase activity, and increasing fructose 2,6 bisphosphate production, which subsequently activates PFK1 (**Figure 1**). Moreover, Glucagon phosphorylates PFK2, activating its phosphatase, which transforms fructose 2,6 bisphosphate back to fructose 6 phosphate. This decreases fructose 2,6 bisphosphate levels and decreases PFK1 activity. Low energy levels within the cell which result in increased AMP and low ATP/AMP ratio, induce allosteric activation of PFK1.

3.1.3 Glycolysis and EC angiogenesis

The endothelium is one of the most diverse tissues in the human body, which displays significant organ-specific heterogeneity. This diversity determines the function of the endothelium according to the organ being supplied [38]. Since ECs lining blood vessels are responsible for supplying oxygen and nutrient to body tissues, the ability to expand this network of blood vessels via angiogenesis is critical for organ growth and function in health and disease [39]. Low oxygen levels serve as a primary stimulus for angiogenesis, which in its classic meaning refers to the sprouting of branches from the existing vessels.

3.1.3.1 Angiogenesis

ECs are essential for the normal functioning of the vascular system. They drive the vascular system expansion during physiologic organ growth to supply sufficient nutrients, as well as under pathologic conditions through a process known as angiogenesis (**Figure 3**). Angiogenesis depends highly on the coordinated orchestra of several regulatory steps [1]. Briefly, this process is guided by the migratory non-proliferative "tip" cells at the forefront from an existing vessel, while the "stalk" cells trail the proliferative and elongation part of the sprout. "Tip" and "stalk" cells continuously switch their phenotype between being either tip or stalk cells. For example, the "tip" cell becomes a "stalk" cell when it loses its migratory behavior, and the "stalk" cell will compete for the position [1]. Several studies found that the vascular endothelial growth factor (VEGF) controls the "tip" cells induction, filopodia formation, and expression of the Notch ligand Delta–like 4 (NLD4) [40, 41]. NLD4, subsequently, suppresses VEGF receptor 2

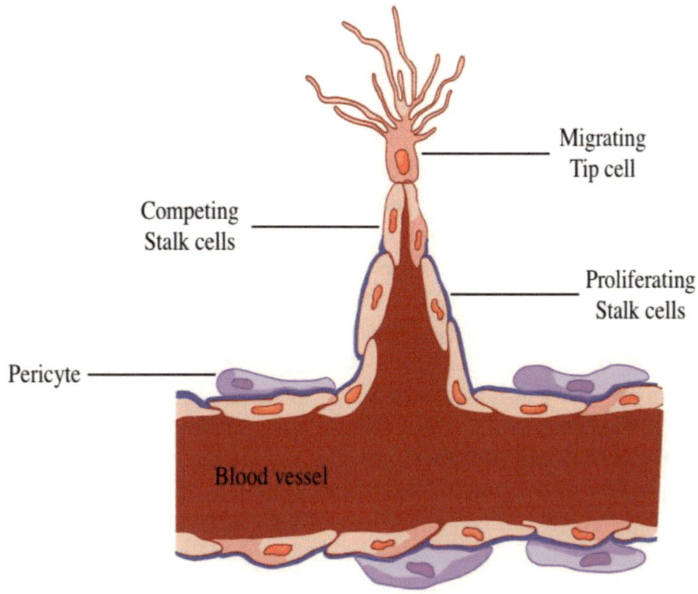

Figure 3.
Angiogenesis is mainly regulated via VEGF. Tip cells require increasing amounts of ATP necessary for migration into hypoxic tissues while proliferating stalk cells generate building blocks (dNTP, protein) to maintain their growth and cellular division.

(VEGFR2/kdr/Flk1) and thus modulates the tip cell behavior. While many genetic and molecular signaling pathways were recognized to be part of this process, the role of ECs metabolism has not been studied and explored until recently.

Switching on the angiogenic machinery of ECs has significant consequences on EC metabolism. This is because angiogenic ECs require nutrients and energy not only for motility but also for the synthesis of building blocks (proteins, nucleotides, and lipids) for cellular proliferation. Hence, during angiogenesis, ECs must increase their metabolic activity to generate energy quickly, while at the same time meeting the challenge of scarce resources as they proliferate in harsh hypoxic environments. Therefore, EC metabolism has to be flexible to support vessel formation under different conditions [39, 42].

Upon switching from quiescence state to vessel branching, the rate of glycolysis is increased in order to fuel subcellular processes required for migration such as cytoskeleton remodeling. Notably, the pro-angiogenic VEGF increases expression of the glycolysis activator phosphofructokinase-2/fructose-2,6-bisphosphatase 3 (PFKFB3) [43]. PFKFB3 generates higher levels of fructose-2,6-bisphosphate, which activates phosphofructokinase 1, the rate limiting enzyme in glycolysis [43, 44]. In fact, studies have shown that genetic and pharmacologic inhibition of the phosphofructokinase 2 reduced EC sprouting and branching capacity [44, 45]. Another regulator enzyme is the hexokinase-2 (HK2) which phosphorylate glucose to glucose-6-phosphate [44, 46]. Several transcription factors such as KLF2 and forkhead box 1 (FOXO1) were found to suppress these key glycolytic enzymes in the quiescent phalanx cells [47, 48]. However, the rate of glycolysis increases in the actively sprouting tip and stalk cells due to VEGF-mediated activation of PFKFB3 and the decreased levels of KLF2 and FOXO1. Interestingly, PFKFB3 and other glycolytic enzymes are highly concentrated in filopodia to generate ATP at the so-called 'ATP hot-spots'. And several studies showed that pharmacologic or genetic inhibition PFKFB3 impairs new vessel formation [43, 44].

3.1.3.2 Endothelial cell metabolism

Despite the fact that oxygen is readily available for EC consumption, the glycolytic pathway remains primary source of energy for EC [38, 44, 49]. In fact, 85% of EC energy production in the form of ATP is generated through glycolysis, even though oxidative phosphorylation (OXPHOS) can generate significantly larger amounts of ATP molecules at much faster rate [43, 49, 50]. However, ECs have fewer mitochondria and consume lower amounts of O2 than other cell types, especially in the presence of abundant supplies of glucose, where only a small fraction of pyruvate is shifted to the TCA cycle [39, 43, 50]. Nonetheless, ECs retain their capacity for oxidative metabolism when glycolysis is compromised or in conditions of stress. Surprisingly, even though the amount of energy generated per glucose molecule via oxidative phosphorylation is significantly greater, higher rates of glycolysis can provide more ATP in a shorter period of time when glucose supply is unlimited. In fact, the rate of glycolysis is high in EC compared to other normal cells, and their glucose consumption is comparable with that of some cancer cells [43].

A logical question that can be asked here is, why do ECs depend on glycolysis for their energy production when they have direct supply of oxygen from blood? There are several explanations for this observation. First, despite the fact that the energy yield via glycolysis is significantly low compared to aerobic respiration, glycolysis can generate ATP molecules at a much faster rate [39, 49]. This is especially important when considering the energy requirements of ECs during angiogenesis. In addition, anaerobic glycolysis facilitates ECs sprouting and proliferation in hypoxic tissues and makes them resistant to hypoxic insults [39]. Also, it limits ROS generation and produces larger amounts of lactic acid, which acts as a pro-angiogenic factor [38, 39, 44]. Moreover, oxygen can be spared to be utilized by the underlying tissue cells. The low oxygen dependence allows sprouting cells to explore and reach distant hypoxic tissues [44]. Also, low oxygen consumption by ECs facilitates oxygen delivery to vital organs. Furthermore, glycolysis provides essential metabolites that are used in multiple cellular pathways such as pentose phosphate pathway (PPP), hexosamine biosynthesis pathway (HBP) and 3-phosphoglycerate (G3P) which generate important molecules and compounds that are used in different cellular processes [39, 49]. Thus, glycolysis provides a metabolic platform that allows ECs to perform diverse roles in the growing and resting vasculature with minimal ROS generation.

3.1.3.3 Alternative metabolism of glucose

ECs engage in several other pathways that can potentially affect angiogenesis, but their exact roles are understudied. Once phosphorylated by hexokinase (HK), glucose-6-phosphate (G6P) can be used to form glycogen, which could serve as an endogenous source of glucose when ECs sprout into glucose-deprived milieu. In fact, inhibition of glycogen phosphorylase (PYG), was found to impair EC migration [51].

G6P can also enter the pentose phosphate pathway (PPP) to generate NADPH [44]. NADPH is essential for restoring the reduced form of glutathione (GSH) from its oxidized form (GSSG), which serves as an antioxidant [38, 52]. PPP provides two intermediates of glycolysis, fructose-6-phosphate (F6P) and glyceraldehyde-3-phosphate (G3P). Interestingly, inhibition of G6P dehydrogenase (G6PD) or Transketolase (TKT) in the PPP was found to impair EC viability and migration [44].

3.1.3.4 Pathways regulation

ECs react to environmental conditions and energy requirements through several mechanisms that involve cellular molecules sensing changes in energy levels. One of these molecules is the AMP-kinase (AMPK), which gets activated by the rising levels of AMP as energy levels dwindle. Activation of AMPK-mediated phosphorylation of metabolic targets promotes catabolic pathways and ATP production, while inhibiting anabolic pathways that consume ATP [39, 53]. This allows ECs to balance their energy level according to environmental changes. For instance, AMPK increases energy production via fatty acid oxidation (FAO) in EC mitochondria and help maintain ATP levels when glucose supplies are low [39, 54]. In addition, AMPK is activated by EC-specific stimuli such as hypoxia and shear stress generated by blood flow [38]. Interestingly, inhibition of AMPK was found to hinder EC angiogenesis in response to hypoxia [55].

3.2 Oxidative phosphorylation

The mechanism by which ATP is produced in the mitochondria via oxidative phosphorylation (OxPhos) was first discovered in the second half of the twentieth century [56, 57]. OxPhos is a process that involves the use of high-energy intermediates for energy transduction between the electron transport chain of the mitochondria and the chemical synthesis of ATP from ADP and phosphate. OxPhos generates 15 times the amount of ATP produced by glycolysis during anaerobic conditions. The reaction involves oxygen consumption, and energy is released from the high energy molecules (NADH, $FADH_2$) and stored in the form of an electrochemical proton gradient across the inner mitochondrial membrane. This energy extraction occurs in three steps each catalyzed by a specific membrane complex including Complex I (NADH dehydrogenase), Complex III (Cytochrome bc_1) and Complex IV (Cytochrome oxidase/COX). Complex II (Succinate dehydrogenase) converts succinate to fumarate, a TCA cycle intermediate, and in the process H^+ is produced from $FADH_2$, which is then shunted by Complex III across the inner mitochondrial membrane. COX is also considered the rate-limiting step of this aerobic respiration. Eventually, the electrochemical proton gradient is utilized by Complex V (ATP Synthase) to produce ATP, or it can be dissipated in the form of heat by passive proton leakage [56, 58, 59].

The electron transport chain is regulated through different mechanisms. Allosteric effectors such as ADP and ATP regulate the process by binding to their specific binding sites on the different mitochondrial complexes. Regulation of the enzyme activity by ATP or ADP binding to the same site on the complex subunit depends on the ATP/ADP ratio. For instance, the exchange of bound ADP by ATP on COX results in an allosteric ATP synthesis inhibition at an ATP/ADP ratio of 28 [60]. In addition, phosphorylation and dephosphorylation of the enzyme complexes is considered another mean of regulating the electron transport chain. For example, phosphorylation of COX was found to inhibit the enzyme activity [56].

3.3 Fatty acid oxidation contribute to dNTP synthesis

Deoxyribonucleoside Triphosphate (dNTP) is a molecule consisting of a deoxyribose sugar attached to three phosphate groups and one of the nucleotide bases, adenine, guanine, cytosine, or thymine as shown in **Figure 4** [61]. Apart from DNA replication, dNTPs may also function as a source of energy for different cellular reactions and signaling pathways [62].

3.3.1 dNTP formation

There are two biosynthetic pathways for nucleotides formation: *de novo* and salvage [62]. The *de novo* pathways require high energy and the use of raw material

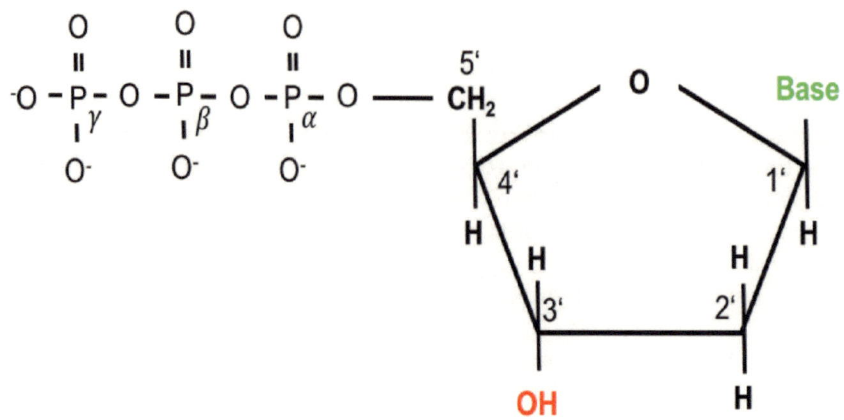

Figure 4.
Structure of Deoxyribonucleoside Triphosphate (dNTP).

like glucose, glutamine, aspartate, and HCO_3 to form nucleotides [62, 63]. However, salvage pathways exist as an alternative energy-efficient route to form nucleotides [63].

The enzyme ribonucleotide reductase (RR), which is NADPH-dependent, is responsible for catalyzing the rate-limiting reaction in which ribonucleotides are converted to their respective deoxyribonucleotides [62, 63]. This reaction is regulated by the number of RR enzymes and allosteric control mechanism [62, 63]. RR consists of two nonidentical subunits, α and β. α subunit has the catalytic site, substrate-specificity site, and activity site; whereas the β subunit contains a stable tyrosyl free radical [63]. The activity of RR enzymes is tightly controlled by allosteric mechanism [63, 64]. The reduction of ribonucleotides requires a specific positive effector, however, the product dNTP can also serve as a negative effector on the enzyme (**Table 1**) [61, 63].

dNTPs levels and RR enzyme activity are important to control the fidelity of nuclear and mitochondrial DNA replication and repair. It has been reported that increased levels of dNTP, *in vitro*, decreased the length of 'S phase' of the cell cycle during DNA replication, which implies that under physiological conditions, nucleotides are used mainly for DNA synthesis [65, 66]. Interestingly, whereas elevated levels of dNTP resulted in delay in the S phase entry through unclear mechanisms [67, 68], depletion of dNTP pool also resulted in inhibition of DNA replication, and fork stalling [69]. In fact, when the enzyme RR was blocked, DNA synthesis was arrested, preserving the dNTP for DNA damage repair under suboptimal conditions [69, 70].

3.3.2 Mitochondrial dNTP

Mitochondria are one of the major endomembrane organelles in eukaryotic cells [14, 71, 72] owing to their ability to produce ATP through oxidative phosphorylation as discussed in Section 3.2. Yet they participate in cellular function and dysfunction, including calcium regulation, activation of cellular death, ROS formation, and cellular building block synthesis [73, 74]. In ECs, the mitochondria compromise only 6% of cell volume, implicating that EC rely on anaerobic glycolysis rather than mitochondria-derived energy [7, 71]. However, mitochondria act primarily as major signaling organelles in the ECs and maintain mitochondrial dNTP pools for proper EC functions. Additionally, alternation in the levels of mitochondrial ROS has been

Substrate	ADP	GDP	CDP	UDP
Positive Effector	dGTP	dTTP	ATP	ATP
Negative Effector	dATP	dATP	dATP dGTP dTTP	dATP dGTP dTTP

Table 1.
Ribonucleotide reductase enzyme regulators.

shown to be associated with impaired one-carbon metabolism, which is essential for purines and pyrimidines nucleotides [75, 76].

In response to mild oxidative stress, the mitochondria attempt to re-establish homeostasis by ROS-buffering capacity of mitochondria. For example, the activity of adenine nucleotide translocase is impaired under mitochondrial oxidation, leading to shortage of adenine diphosphate (ADP) [77]. On the other hand, up-regulation of mitochondrial anti-oxidant systems and other molecules counteract ROS-induced protein unfolding [78, 79]. If oxidative stress is persistent, mitochondria may translate the adaptive response into activation of cellular death [74]. These responses and deregulation of ROS levels contribute to the pathogenesis of cardiovascular system, including coronary artery diseases.

3.3.3 Fatty acid oxidation (FAO)

Long chain fatty acids are a major source of energy productions, primarily in mitochondria [61, 80, 81]. Fatty acids are broken up into acetyl CoA, NADH and $FADH_2$ in the mitochondria [80]. These three products are used by the mitochondrial matrix for energy production through TCA and oxidative phosphorylation [80].

3.3.3.1 Fatty acid oxidation as a major energy-producing pathway

Fatty acid oxidation (FAO) is an important catabolic and anabolic process. On the outer membrane of mitochondria, FAO transfers the acyl group from CoA to carnitine by carnitine palmitoyltransferase I (CPT1). Acyl-carnitine is then exchanged across the inner membrane of mitochondria. The acyl group is transferred back again to CoA by carnitine palmitoyltransferase II (CPT II) as shown in **Figure 5** [82]. CPT1 is an important enzyme for FAO and is a rate limiting factor for FAO in the mitochondria. Malonyl CoA, an intermediate product of fatty acid synthesis, is an inhibitor of CPT1.

β-oxidation is a four steps process carried by enzymatic oxidation, hydration, and oxidation that act on acyl CoA to yield a shorter acyl CoA and acetyl CoA [83]. The four-step process is shown in the schematics of **Figure 6**.

3.3.3.2 Role of fatty acid oxidation in vessel sprouting

Recent studies have shown the critical role of FAO for vessel sprouting [42, 84]. In a study, the levels of FAO and dNTP synthesis were reduced when mitochondrial CPT1A was silenced. This resulted in impaired vascular sprouting due to reduction ECs proliferation but not migration. Additionally, silencing long-chain acyl-CoA dehydrogenase (ACADVL) has yielded similar results, supporting the role of FAO in vessel sprouting. Overexpression of CPT1A obtained opposite results, further supporting a crucial role of CPT1A in angiogenesis.

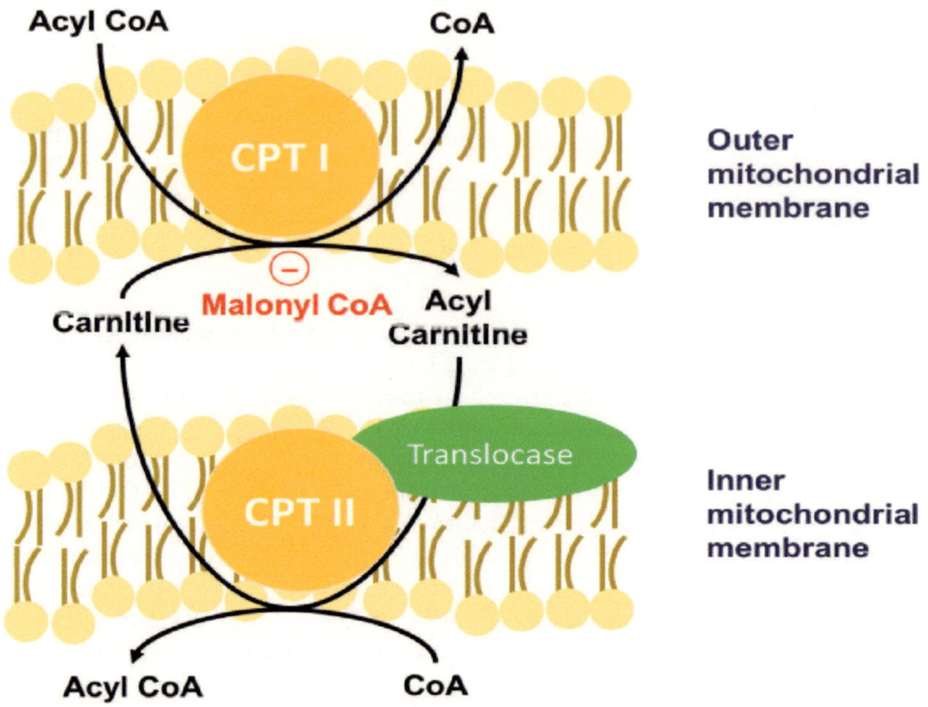

Figure 5.
Long-chain fatty acid transportation in the mitochondra. Fatty acids are transported through the mitochondrial membrane as acyl CoA for subsequent oxidation. Malonyl CoA acts as a key inhibitor molecule for CPT I, and thus regulating the rate of fatty acid oxidation (FAO).

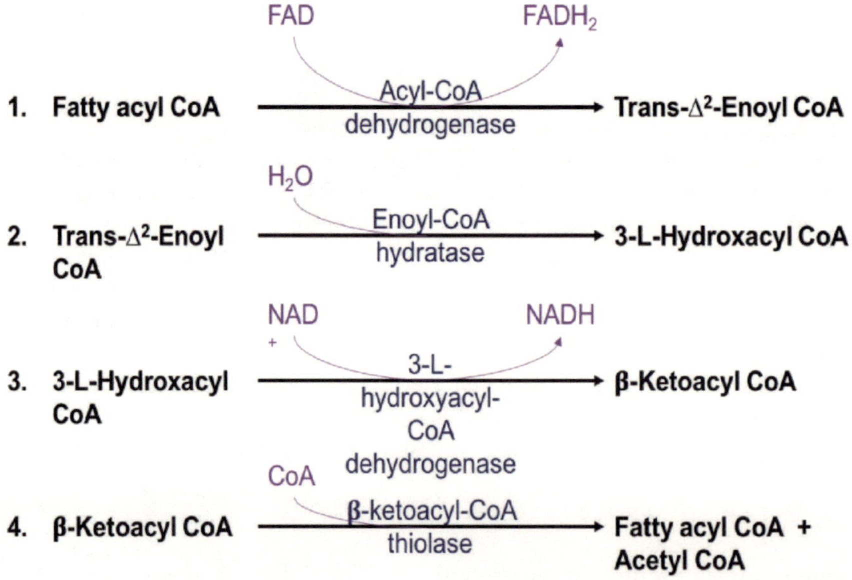

Figure 6.
Fatty acid β-oxidation pathway.

3.3.3.3 Fatty acid oxidation for de novo *synthesis of nucleotides*

As noted above, silencing CPT1A in ECs showed impaired *de novo* synthesis of dNTPs. This impaired *de novo* dNTP synthesis contributed to reduced vessel sprouting [82]. Nonetheless, the levels of glucose oxidation were increased to compensate for the FAO loss, yet it was not sufficient to help in the proliferative defects of ECs with knockdown CPT1A. This reflects the irreplaceable role of FAO for *de novo* dNTP synthesis in ECs [82].

3.3.3.4 Fatty acid β-oxidation in quiescent vs. proliferating endothelial cells

Depending on the cellular status, the FAO are directed either toward DNA synthesis or redox homeostasis. FAO are involved in regenerating $NADP^+$ to NADPH, they also upregulate the expression of $NADP^+$ producing genes, which are critical for redox homeostasis [43, 82]. Quiescent ECs upregulate FAO, but do not rely on them for ATP production or nucleotide synthesis, rather utilize it for redox homeostasis [85]. Unlike quiescent ECs, proliferating ECs utilize FAO for DNA synthesis, as previously discussed [82].

CPT1A, the rate limiting enzyme for FAO in mitochondria, has been shown to be critical for redox homeostasis in EC. In quiescent ECs, CPT1A inhibition caused the levels of ROS to elevate, leading to decreased anti-fibrinolytic gene expression, endothelial leakage, and increased leukocytes adhesion and/or infiltration [80, 85]. Thus, it is believed that quiescent ECs require more redox buffering capacity compared to proliferating ECs due to higher levels of ROS.

Besides the involvement of FAO in redox balance in quiescent ECs, they are also involved in other vasculo-protective NADPH-regenerating pathways such as oxidative PPP and nicotine nucleotide transhydrogenase [85].

4. Endothelial metabolism in atherosclerosis

The generation of increased amounts of NO in atherosclerosis is critical for its anti-atherogenic effects, including vasodilation, inhibition of platelet aggregation, smooth muscle proliferation as well as leukocyte migration and oxidative stress [38]. Endothelial cells produce NO through enzymatic oxidation of arginine to citrulline via eNOS enzyme. eNOS requires several co-factors including NADPH, flavin adenine dineucleotide (FAD), flavin mononeucleotide (FMN), Calcium/Calmodulin and tetrahydrobiopterin (BH4). Decreased availability of Arginine or deficiency of BH4 results in the paradoxical generation of ROS instead of NO by eNOS, a process known as eNOS uncoupling [38, 86]. Arginine in particular, has been found to be rate-limiting for NO synthesis in patients with atherosclerosis [87]. It was demonstrated that an arginine analog asymmetric dimethyl arginine (ADMA), that acts as a competitor for eNOS, impaired NO production. ADMA levels are markedly increased in atherosclerosis and therefore it is recognized as a major cardiovascular risk factor [88]. Moreover, Dimethyl arginine dimethyl aminohydrolase (DDAH), an enzyme that metabolizes ADMA into citrulline and dimethylamine is impaired by the oxidative stress in atherosclerosis [38]. Interestingly, because of this competition, Arginine supplements have been found to be of great benefit in atherosclerotic patients with high ADMA levels, by enhancing endothelial-dependent vasodilation and inhibition of leukocyte adhesion and migration to the atherosclerotic plaque [89].

Furthermore, endothelial NADPH oxidase is induced by certain atherosclerotic plaque components such as the oxidized LDL (oxLDL). The NADPH oxidase-derived ROS were found to have detrimental effects in promoting plaque progression. These include oxidation of LDLs, inducing vascular smooth muscle proliferation and migration and EC apoptosis as well as promoting the expression of vascular adhesion adhesion molecules [38].

5. Conclusions

The endothelium is one of the most diverse tissues in the human body. It maintains the integrity of the vascular system and provides nutrition to underlying tissues. In addition, EC drives the growth and proliferation of blood vessels under physiologic and pathologic conditions. ECs exhibit significant flexibility in response to various environmental changes such as hypoxia and ischemia. Careful analysis of the process of sprouting angiogenesis explains how ECs function in such an orchestrated way to reach their end goal of providing nutrients and oxygen supply to the affected tissues. ECs display phenomenal resilience in the process through various mechanisms, one of which is their metabolic adaptation and the other is critical balance between subcellular levels of ROS (cytosolic versus mitochondrial). ECs limit their oxygen consumption in order to preserve it for the tissues that they supply to and also to maintain a balanced intracellular redox state. Although ECs do not utilize mitochondrial OxPhos for ATP synthesis and thus generate very little mitochondrial ROS, NADPH oxidase-derived ROS appear to regulate many critical EC functions in health and disease. However, EC has intricated intracellular mechanisms by which subcellular oxidants may communicate at the subcellular levels [7]. Unlike most cells in the body (except tumor cells), ECs upregulate and accelerate their glycolytic pathways in order to generate energy (ATP production) and certain molecules that act as building blocks (dNTPs) and are essential for supporting EC proliferation and migration. CPT1A-mediated FAO appears to play a significant role in synthesizing dNTPs and $NADP^+$, NADPH in EC mitochondria. Depending on the metabolic states of ECs (quiescent versus proliferative), FAO-generated NADPH is utilized for quiescent EC's redox homeostasis or dNTPS for cell proliferation in angiogenic endothelium. Further studies aimed at understanding the molecular mechanisms by which subcellular ROS modulate EC metabolism in health and disease will help develop therapeutics modalities for CVD.

Acknowledgements

This work was supported by the National Heart, Lung, and Blood Institute (NHLBI) 1R01HL133624 (M.R.A.); R01HL46716 and R01HL128831-01A1 (F.W.S).

Conflict of interest

The authors declare no conflict of interest.

Author details

Sarah R. Aldosari[1], Maan A. Awad[1], Frank W. Sellke[2] and Md. Ruhul Abid[2]*

1 College of Medicine, Alfaisal University, Riyadh, Saudi Arabia

2 Cardiovascular Research Center, Cardiothoracic Surgery Division, Rhode Island Hospital, Brown University Warren Alpert Medical School, Providence, RI, USA

*Address all correspondence to: ruhul_abid@brown.edu

References

[1] Vandekeere S, Dewerchin M, Carmeliet P. Angiogenesis revisited: An overlooked role of endothelial cell metabolism in vessel sprouting. Microcirculation. 2015;**22**(7):509-517. DOI: 10.1111/micc.12229

[2] Cantelmo AR, Brajic A, Carmeliet P. Endothelial metabolism driving angiogenesis: Emerging concepts and principles. The Cancer Journal. 2015;**21**(4):244-249. DOI: 10.1097/PPO.0000000000000133

[3] Eelen G, Cruys B, Welti J, De Bock K, Carmeliet P. Control of vessel sprouting by genetic and metabolic determinants. Trends in Endocrinology and Metabolism: TEM. 2013;**24**(12):589-596. DOI: 10.1016/j.tem.2013.08.006

[4] Ruiz-Lozano P, Rajan P. Stem cells as in vitro models of disease. Current Stem Cell Research & Therapy. 2007;**2**(4):280-292. DOI: 10.2174/157488807782793772

[5] Fearon IM, Gaça MD, Nordskog BK. In vitro models for assessing the potential cardiovascular disease risk associated with cigarette smoking. oxicology in Vitro: An International Journal Published in Association with BIBRA. 2013;*27*(1):513-522. DOI: 10.1016/j.tiv.2012.08.018

[6] Shafique E, Choy WC, Liu Y, Feng J, Cordeiro B, Lyra A, et al. Oxidative stress improves coronary endothelial function through activation of the pro-survival kinase AMPK. Aging. 2013;**5**(7):515-530 https://doi.org/10.18632/aging.100569

[7] Shafique E, Torina A, Reichert K, Colantuono B, Nur N, Zeeshan K, et al. Mitochondrial redox plays a critical role in the paradoxical effects of NAPDH oxidase-derived ROS on coronary endothelium. Cardiovascular Research. 2017;**113**(2):234-246. DOI: 10.1093/cvr/cvw249

[8] Aldosari S, Awad M, Harrington EO, Sellke FW, Abid MR. Subcellular reactive oxygen species (ROS) in cardiovascular pathophysiology. Antioxidants. 2018;**7**(1):14. DOI: 10.3390/antiox7010014

[9] Kim YM, Kim SJ, Tatsunami R, Yamamura H, Fukai T, Ushio-Fukai M. ROS-induced ROS release orchestrated by Nox4, Nox2, and mitochondria in VEGF signaling and angiogenesis. American Journal of Physiology. Cell Physiology. 2017;**312**(6):C749-C764. DOI: 10.1152/ajpcell.00346.2016

[10] Schröder K, Zhang M, Benkhoff S, Mieth A, Pliquett R, Kosowski J, et al. Nox4 is a protective reactive oxygen species generating vascular NADPH oxidase. Circulation Research. 2012;**110**(9):1217-1225. DOI: 10.1161/CIRCRESAHA.112.267054

[11] Awad MA, Aldosari SR, Abid MR. Genetic alterations in oxidant and anti-oxidant enzymes in the vascular system. Frontiers in Cardiovascular Medicine. 2018;**5**:107. DOI: 10.3389/fcvm.2018.00107

[12] Sleight P. The HOPE study (heart outcomes prevention evaluation). Journal of the Renin-Angiotensin-Aldosterone System. 2000;**1**(1):18-20. DOI: 10.3317/jraas.2000.002

[13] Ponnuswamy P, Schröttle A, Ostermeier E, Grüner S, Huang PL, Ertl G, et al. eNOS protects from atherosclerosis despite relevant superoxide production by the enzyme in apoE mice. PLoS One. 2012;**7**(1):e30193. DOI: 10.1371/journal.pone.0030193

[14] Aldosari S, Awad M, Gao MZ, McCormack IG, Sellke FW, Abid MR. Oxidant-dependent and oxidant-independent proangiogenic and vasomotor signaling in coronary vascular endothelium. In: Chakraborti S,

Dhalla N, Dikshit M, Ganguly N, editors. Modulation of Oxidative Stress in Heart Disease. Singapore: Springer; 2019

[15] Frey RS, Ushio-Fukai M, Malik AB. NADPH oxidase-dependent signaling in endothelial cells: Role in physiology and pathophysiology. Antioxidants & Redox Signaling. 2009;**11**(4):791-810. DOI: 10.1089/ars.2008.2220

[16] Manuneedhi Cholan P, Cartland SP, Kavurma MM. NADPH oxidases, angiogenesis, and peripheral artery disease. Antioxidants. 2017;**6**(3):56. DOI: 10.3390/antiox6030056

[17] Bedard K, Krause KH. The NOX family of ROS-generating NADPH oxidases: Physiology and pathophysiology. Physiological Reviews. 2007;**87**(1):245-313. DOI: 10.1152/physrev.00044.2005

[18] Feng J, Damrauer SM, Lee M, Sellke FW, Ferran C, Abid MR. Endothelium-dependent coronary vasodilatation requires NADPH oxidase-derived reactive oxygen species. Arteriosclerosis, Thrombosis, and Vascular Biology. 2010;**30**(9):1703-1710. DOI: 10.1161/ATVBAHA.110.209726

[19] Szöcs K, Lassègue B, Sorescu D, Hilenski LL, Valppu L, Couse TL, et al. Upregulation of Nox-based NAD(P)H oxidases in restenosis after carotid injury. Arteriosclerosis, Thrombosis, and Vascular Biology. 2002;**22**(1):21-27. DOI: 10.1161/hq0102.102189

[20] Schröder K, Helmcke I, Palfi K, Krause KH, Busse R, Brandes RP. Nox1 mediates basic fibroblast growth factor-induced migration of vascular smooth muscle cells. Arteriosclerosis, Thrombosis, and Vascular Biology. 2007;**27**(8):1736-1743. DOI: 10.1161/ATVBAHA.107.142117

[21] Sheehan AL, Carrell S, Johnson B, Stanic B, Banfi B, Miller FJ Jr. Role for Nox1 NADPH oxidase in atherosclerosis. Atherosclerosis. 2011;**216**(2):321-326. DOI: 10.1016/j.atherosclerosis.2011.02.028

[22] Mollnau H, Wendt M, Szöcs K, Lassègue B, Schulz E, Oelze M, et al. Effects of angiotensin II infusion on the expression and function of NAD(P)H oxidase and components of nitric oxide/cGMP signaling. Circulation Research. 2002;**90**(4):E58-E65. DOI: 10.1161/01.res.0000012569.55432.02

[23] Gavazzi G, Banfi B, Deffert C, Fiette L, Schappi M, Herrmann F, et al. Decreased blood pressure in NOX1-deficient mice. FEBS Letters. 2006;**580**(2):497-504. DOI: 10.1016/j.febslet.2005.12.049

[24] Guzik TJ, Chen W, Gongora MC, Guzik B, Lob HE, Mangalat D, et al. Calcium-dependent NOX5 nicotinamide adenine dinucleotide phosphate oxidase contributes to vascular oxidative stress in human coronary artery disease. Journal of the American College of Cardiology. 2008;**52**(22):1803-1809. DOI: 10.1016/j.jacc.2008.07.063

[25] Bendall JK, Rinze R, Adlam D, Tatham AL, de Bono J, Wilson N, et al. Endothelial Nox2 overexpression potentiates vascular oxidative stress and hemodynamic response to angiotensin II: Studies in endothelial-targeted Nox2 transgenic mice. Circulation Research. 2007;**100**(7):1016-1025. DOI: 10.1161/01.RES.0000263381.83835.7b

[26] Chen L, Hou X, Xiao J, Kuroda J, Ago T, Sadoshima J, et al. Both hydrogen peroxide and transforming growth factor beta 1 contribute to endothelial Nox4 mediated angiogenesis in endothelial Nox4 transgenic mouse lines. Biochimica et Biophysica Acta (BBA) - Molecular Basis of Disease. 2014;**1842**(12):2489-2499. DOI: 10.1016/j.bbadis.2014.10.007

[27] Datla SR, Peshavariya H, Dusting GJ, Mahadev K, Goldstein BJ,

Jiang F. Important role of Nox4 type NADPH oxidase in angiogenic responses in human microvascular endothelial cells in vitro. Arteriosclerosis, Thrombosis, and Vascular Biology. 2007;**27**(11):2319-2324. DOI: 10.1161/ATVBAHA.107.149450

[28] Ray R, Murdoch CE, Wang M, Santos CX, Zhang M, Alom-Ruiz S, et al. Endothelial Nox4 NADPH oxidase enhances vasodilatation and reduces blood pressure in vivo. Arteriosclerosis, Thrombosis, and Vascular Biology. 2011;**31**(6):1368-1376. DOI: 10.1161/ATVBAHA.110.219238

[29] Craige SM, Chen K, Pei Y, Li C, Huang X, Chen C, et al. NADPH oxidase 4 promotes endothelial angiogenesis through endothelial nitric oxide synthase activation. Circulation. 2011;**124**(6):731-740. DOI: 10.1161/CIRCULATIONAHA.111.030775

[30] Shimokawa H, Morikawa K. Hydrogen peroxide is an endothelium-derived hyperpolarizing factor in animals and humans. Journal of Molecular and Cellular Cardiology. 2005;**39**(5):725-732. DOI: 10.1016/j.yjmcc.2005.07.007

[31] Matoba T, Shimokawa H, Nakashima M, Hirakawa Y, Mukai Y, Hirano K, et al. Hydrogen peroxide is an endothelium-derived hyperpolarizing factor in mice. The Journal of Clinical Investigation. 2000;**106**(12):1521-1530. DOI: 10.1172/JCI10506

[32] Miura H, Bosnjak JJ, Ning G, Saito T, Miura M, Gutterman DD. Role for hydrogen peroxide in flow-induced dilation of human coronary arterioles. Circulation Research. 2003;**92**(2):e31-e40. DOI: 10.1161/01.res.0000054200.44505.ab

[33] BelAiba RS, Djordjevic T, Petry A, Diemer K, Bonello S, Banfi B, et al. NOX5 variants are functionally active in endothelial cells. Free Radical Biology & Medicine. 2007;**42**(4):446-459. DOI: 10.1016/j.freeradbiomed.2006.10.054

[34] Herkert O, Djordjevic T, BelAiba RS, Görlach A. Insights into the redox control of blood coagulation: Role of vascular NADPH oxidase-derived reactive oxygen species in the thrombogenic cycle. Antioxidants and Redox Signaling. 2004;**6**(4):765-776. DOI: 10.1089/1523086041361695

[35] Chaudhry R, Varacallo M. Biochemistry, Glycolysis. StatPearls. 2018. Available from: http://www.ncbi.nlm.nih.gov/pubmed/29493928

[36] Naifeh J, Varacallo M. Biochemistry, Aerobic Glycolysis. StatPearls. 2019. Available from: http://www.ncbi.nlm.nih.gov/pubmed/29262043

[37] Jois T, Sleeman MW. The regulation and role of carbohydrate response element-binding protein in metabolic homeostasis and disease. Journal of Neuroendocrinology. 2017;**29**(10). DOI: 10.1111/jne.12473

[38] Eelen G, de Zeeuw P, Simons M, Carmeliet P. Endothelial cell metabolism in normal and diseased vasculature. Circulation Research. 2015;**116**(7):1231-1244. DOI: 10.1161/CIRCRESAHA.116.302855

[39] Potente M, Carmeliet P. The link between angiogenesis and endothelial metabolism. Annual Review of Physiology. 2017;**79**(1):43-66. DOI: 10.1146/annurev-physiol-021115-105134

[40] Potente M, Gerhardt H, Carmeliet P. Basic and therapeutic aspects of angiogenesis. Cell. 2011;**146**(6):873-887. DOI: 10.1016/j.cell.2011.08.039

[41] Jakobsson L, Franco CA, Bentley K, Collins RT, Ponsioen B, Aspalter IM, et al. Endothelial cells dynamically compete for the tip cell position during angiogenic sprouting. Nature Cell Biology. 2010;**12**(10):943-953. DOI: 10.1038/ncb2103

[42] De Bock K, Georgiadou M, Carmeliet P. Role of endothelial cell metabolism in vessel sprouting. Cell Metabolism. 2013;**18**(5):634-647. DOI: 10.1016/j.cmet.2013.08.001

[43] De Bock K, Georgiadou M, Schoors S, Kuchnio A, Wong BW, Cantelmo AR, et al. Role of PFKFB3-driven glycolysis in vessel sprouting. Cell. 2013;**154**(3):651-663. DOI: 10.1016/j.cell.2013.06.037

[44] Rohlenova K, Veys K, Miranda-Santos I, De Bock K, Carmeliet P. Endothelial cell metabolism in health and disease. Trends in Cell Biology. 2018;**28**(3):224-236. DOI: 10.1016/j.tcb.2017.10.010

[45] Schoors S, De Bock K, Cantelmo AR, Georgiadou M, Ghesquière B, Cauwenberghs S, et al. Partial and transient reduction of glycolysis by PFKFB3 blockade reduces pathological angiogenesis. Cell Metabolism. 2014;**19**(1):37-48. DOI: 10.1016/j.cmet.2013.11.008

[46] Yu P, Wilhelm K, Dubrac A, Tung JK, Alves TC, Fang JS, et al. FGF-dependent metabolic control of vascular development. Nature. 2017;**545**(7653):224-241. DOI: 10.1038/nature22322

[47] Doddaballapur A, Michalik KM, Manavski Y, Lucas T, Houtkooper RH, You X, et al. Laminar shear stress inhibits endothelial cell metabolism via KLF2-mediated repression of PFKFB3. Arteriosclerosis, Thrombosis, and Vascular Biology. 2015;**35**(1):137-145. DOI: 10.1161/ATVBAHA.114.304277

[48] Wilhelm K, Happel K, Eelen G, Schoors S, Oellerich MF, Lim R, et al. FOXO1 couples metabolic activity and growth state in the vascular endothelium. Nature. 2016;**529**(7585):216-220. DOI: 10.1038/nature16498

[49] Heiden MGV, Cantley LC, Thompson CB. Understanding the Warburg effect: The metabolic requirements of cell proliferation. Science. 2009;**324**(5930):1029-1033. DOI: 10.1126/science.1160809

[50] Ghesquière B, Wong BW, Kuchnio A, Carmeliet P. Metabolism of stromal and immune cells in health and disease. Nature. 2014;**511**:167-176. DOI: 10.1038/nature13312

[51] Vizán P, Sánchez-Tena S, Alcarraz-Vizán G, Soler M, Messeguer R, Pujol MD, et al. Characterization of the metabolic changes underlying growth factor angiogenic activation: Identification of new potential therapeutic targets. Carcinogenesis. 2009;**30**(6):946-952. DOI: 10.1093/carcin/bgp083

[52] Riganti C, Gazzano E, Polimeni M, Aldieri E, Ghigo D. The pentose phosphate pathway: An antioxidant defense and a crossroad in tumor cell fate. Free Radical Biology & Medicine. 2012;**53**(3):421-436. DOI: 10.1016/j.freeradbiomed.2012.05.006

[53] Hardie DG, Schaffer BE, Brunet A. AMPK: An energy-sensing pathway with multiple inputs and outputs. Trends in Cell Biology. 2016;**26**(3):190-201. DOI: 10.1016/j.tcb.2015.10.013

[54] Dagher Z, Ruderman N, Tornheim K, Ido Y. The effect of AMP-activated protein kinase and its activator AICAR on the metabolism of human umbilical vein endothelial cells. Biochemical and Biophysical Research Communications. 1999;**265**(1):112-115. DOI: 10.1006/bbrc.1999.1635

[55] Nagata D, Mogi M, Walsh K. AMP-activated protein kinase (AMPK) signaling in endothelial cells is essential for angiogenesis in response to hypoxic stress. Journal of Biological Chemistry. 2003;**278**(33):31000-31006. DOI: 10.1074/jbc.M300643200

[56] Kadenbach B. Mitochondrial Oxidative Phosphorylation: Nuclear-Encoded Genes, Enzyme Regulation, and Pathophysiology. New York: Springer; 2012

[57] Boyer PD. Conformational coupling in oxidative phosphorylation and photophosphorylation. Trends in Biochemical Sciences. 1977;**2**(2):38-41. DOI: 10.1016/0968-0004(77)90254-7

[58] Rees DM, Leslie AG, Walker JE. The structure of the membrane extrinsic region of bovine ATP synthase. Proceedings of the National Academy of Sciences of the United States of America. 2009;**106**(51):21597-21601. DOI: 10.1073/pnas.0910365106

[59] Jastroch M, Divakaruni AS, Mookerjee S, Treberg JR, Brand MD. Mitochondrial proton and electron leaks. Essays in Biochemistry. 2010;**47**:53-67. DOI: 10.1042/bse0470053

[60] Arnold S, Kadenbach B. Cell respiration is controlled by ATP, an allosteric inhibitor of cytochrome-c oxidase. European Journal of Biochemistry. 1997;**249**(1):350-354. DOI: 10.1111/j.1432-1033.1997.t01-1-00350.x

[61] Devlin TM. Textbook of Biochemistry with Clinical Correlation. New York: John Wiley and Sons; 1982

[62] Lane AN, Fan TW. Regulation of mammalian nucleotide metabolism and biosynthesis. Nucleic Acids Research. 2015;**43**(4):2466-2485. DOI: 10.1093/nar/gkv047

[63] Nordlund P, Reichard P. Ribonucleotide reductases. Annual Review of Biochemistry. 2006;**75**:681-706. DOI: 10.1146/annurev.biochem.75.103004.142443

[64] Fairman JW, Wijerathna SR, Ahmad MF, Xu H, Nakano R, Jha S, et al. Structural basis for allosteric regulation of human ribonucleotide reductase by nucleotide-induced oligomerization. Nature Structural and Molecular Biology. 2011;**18**(3):316-322. DOI: 10.1038/nsmb.2007

[65] Stodola JL, Burgers PM. Resolving individual steps of Okazaki-fragment maturation at a millisecond timescale. Nature Structural & Molecular Biology. 2016;**23**(5):402-408. DOI: 10.1038/nsmb.3207

[66] Kunkel TA, Sabatino RD, Bambara RA. Exonucleolytic proof reading by calf thymus DNA polymerase delta. Proceedings of the National Academy of Sciences of the United States of America. 1987;**84**(14):4865-4869. DOI: 10.1073/pnas.84.14.4865

[67] Chabes A, Stillman B. Constitutively high dNTP concentration inhibits cell cycle progression and the DNA damage checkpoint in yeast Saccharomyces cerevisiae. Proceedings of the National Academy of Sciences of the United States of America. 2007;**104**(4):1183-1188. DOI: 10.1073/pnas.0610585104

[68] Franzolin E, Pontarin G, Rampazzo C, Miazzi C, Ferraro P, Palumbo E, et al. The deoxynucleotide triphosphohydrolase SAMHD1 is a major regulator of DNA precursor pools in mammalian cells. Proceedings of the National Academy of Sciences of the United States of America. 2013;**110**(35):14272-14277. DOI: 10.1073/pnas.1312033110

[69] Koç A, Wheeler LJ, Mathews CK, Merrill GF. Hydroxyurea arrests DNA replication by a mechanism that preserves basal dNTP pools. The Journal of Biological Chemistry. 2004;**279**(1):223-230. DOI: 10.1074/jbc.M303952200

[70] Aye Y, Li M, Long MJ, Weiss RS. Ribonucleotide reductase and cancer: Biological mechanisms and targeted

therapies. Oncogene. 2015;**34**(16): 2011-2021. DOI: 10.1038/onc.2014.155

[71] Friedman JR, Nunnari J. Mitochondrial form and function. Nature. 2014;**505**(7483):335-343. DOI: 10.1038/nature12985

[72] Dromparis P, Michelakis ED. Mitochondria in vascular health and disease. Annual Review of Physiology. 2013;**75**:95-126. DOI: 10.1146/annurev-physiol-030212-183804

[73] Osellame LD, Blacker TS, Duchen MR. Cellular and molecular mechanisms of mitochondrial function. Best practice & research. The Journal of Clinical Endocrinology and Metabolism. 2012;**26**(6):711-723. DOI: 10.1016/j.beem.2012.05.003

[74] Galluzzi L, Kepp O, Trojel-Hansen C, Kroemer G. Mitochondrial control of cellular life, stress, and death. Circulation Research. 2012;**111**(9):1198-1207. DOI: 10.1161/CIRCRESAHA.112.268946

[75] Bao XR, Ong SE, Goldberger O, Peng J, Sharma R, Thompson DA, et al. Mitochondrial dysfunction remodels one-carbon metabolism in human cells. eLife. 2016;**5**:e10575. DOI: 10.7554/eLife.10575

[76] Nikkanen J, Forsström S, Euro L, Paetau I, Kohnz RA, Wang L, et al. Mitochondrial DNA replication defects disturb cellular dNTP pools and remodel one-carbon metabolism. Cell Metabolism. 2016;**23**(4):635-648. DOI: 10.1016/j.cmet.2016.01.019

[77] Pestana CR, Silva CH, Uyemura SA, Santos AC, Curti C. Impact of adenosine nucleotide translocase (ANT) proline isomerization on Ca2+-induced cysteine relative mobility/mitochondrial permeability transition pore. Journal of Bioenergetics and Biomembranes. 2010;**42**(4):329-335. DOI: 10.1007/s10863-010-9297-4

[78] Sharma S, Dewald O, Adrogue J, Salazar RL, Razeghi P, Crapo JD, et al. Induction of antioxidant gene expression in a mouse model of ischemic cardiomyopathy is dependent on reactive oxygen species. Free Radical Biology & Medicine. 2006;**40**(12):2223-2231. DOI: 10.1016/j.freeradbiomed.2006.02.019

[79] Singh M, Sharma H, Singh N. Hydrogen peroxide induces apoptosis in HeLa cells through mitochondrial pathway. Mitochondrion. 2007;**7**(6):367-373. DOI: 10.1016/j.mito.2007.07.003

[80] Patella F, Schug ZT, Persi E, Neilson LJ, Erami Z, Avanzato D, et al. Proteomics-based metabolic modeling reveals that fatty acid oxidation (FAO) controls endothelial cell (EC) permeability. Molecular & Cellular Proteomics. 2015;**14**(3):621-634. DOI: 10.1074/mcp.M114.045575

[81] Dagher Z, Ruderman N, Tornheim K, Ido Y. Acute regulation of fatty acid oxidation and amp-activated protein kinase in human umbilical vein endothelial cells. Circulation Research. 2001;**88**(12):1276-1282. DOI: 10.1161/hh1201.092998

[82] Schoors S, Bruning U, Missiaen R, Queiroz KC, Borgers G, Elia I, et al. Fatty acid carbon is essential for dNTP synthesis in endothelial cells. Nature. 2015;**520**(7546):192-197. DOI: 10.1038/nature14362

[83] Hülsmann WC, Dubelaar ML. Aspects of fatty acid metabolism in vascular endothelial cells. Biochimie. 1988;**70**(5):681-686. DOI: 10.1016/0300-9084(88)90253-2

[84] Teuwen LA, Draoui N, Dubois C, Carmeliet P. Endothelial cell metabolism: An update anno 2017. Current Opinion in Hematology. 2017;**24**(3):240-247. DOI: 10.1097/MOH.0000000000000335

[85] Kalucka J, Bierhansl L, Conchinha NV, Missiaen R, Elia I, Brüning U, et al. Quiescent endothelial cells upregulate fatty acid β-oxidation for vasculoprotection via redox homeostasis. Cell Metabolism. 2018;**28**(6):881-894.e13. DOI: 10.1016/j.cmet.2018.07.016

[86] Kawashima S. Malfunction of vascular control in lifestyle-related diseases: Endothelial nitric oxide (NO) synthase/NO system in atherosclerosis. Journal of Pharmacological Sciences. 2004;**96**(4):411-419. DOI: 10.1254/jphs.fmj04006x6

[87] Böger RH, Bode-Böger SM, Szuba A, Tsao PS, Chan JR, Tangphao O, et al. Asymmetric dimethylarginine (ADMA): A novel risk factor for endothelial dysfunction - Its role in hypercholesterolemia. Circulation. 1998;**98**(18):1842-1847. DOI: 10.1161/01.CIR.98.18.1842

[88] Böger RH. Asymmetric dimethylarginine: Understanding the physiology, genetics, and clinical relevance of this novel biomarker. In: Proceedings of the 4th International Symposium on ADMA. Pharmacological Research; 2009. DOI: 10.1016/j.phrs.2009.10.001

[89] Chan JR, Böger RH, Bode-Böger SM, Tangphao O, Tsao PS, Blaschke TF, et al. Asymmetric dimethylarginine increases mononuclear cell adhesiveness in hypercholesterolemic humans. Arteriosclerosis, Thrombosis, and Vascular Biology. 2000;**20**(4):1040-1046. DOI: 10.1161/01.ATV.20.4.1040

Section 2

Oxidoreductase as a Therapeutic Target

Chapter 4

The Impact of Oxidoreductases-Related MicroRNAs in Glucose Metabolism of Renal Cell Carcinoma and Prostate Cancer

Mariana Gomes Morais, Francisca Guilherme Carvalho Dias, João Alexandre Velho Prior, Ana Luísa Pereira Teixeira and Rui Manuel de Medeiros Melo Silva

Abstract

The reprogramming of metabolism is one of cancer hallmarks. Glucose's metabolism, as one of the main fuels of cancer cells, has been the focus of several research studies in the oncology field. However, because cancer is a heterogeneous disease, the disruptions in glucose metabolism are highly variable depending of the cancer. In fact, Renal Cell Carcinoma (RCC) and Prostate Cancer (PCa), the most lethal and common urological neoplasia, respectively, show different disruptions in the main pathways of glucose catabolism: glycolysis, lactate fermentation and Krebs Cycle. Oxidoreductases are a class of enzymes that catalyze electrons transfer from one molecule to another and are present in these three pathways, posing as an opportunity to better understand these catabolic deregulations. Furthermore, nowadays it is recognized that their expression is modulated by microRNAs (miRNAs), in this book chapter, we selected the known miRNAs that directly target these oxidoreductases and analyzed their deregulation in both cancers. The characterization of these miRNAs opens a new door that could be applied in patients' stratification and therapy monitorization because of their potential as cancer biomarkers. Additionally, their delivery to cancer cells, using glucose capped NPs could help establish new therapeutic strategies that would improve RCC and PCa management.

Keywords: oxidoreductases, urological cancers, glucose metabolism, biomarkers, therapeutic targets, nanoparticles

1. Introduction

Cancer is one of the current main public health problems in the world, accounting for, according to GLOBOCAN, approximately 18.1 million new cases and 9.6 million deaths, worldwide in 2018 [1]. It arises from genetic and environmental interactions that cause the deregulation of signaling pathways involved in fundamental cellular processes. Being a heterogeneous disease with multiple etiologies, cancer shows different pathological evolutions and treatment approaches [2].

Renal Cell Carcinoma (RCC) and Prostate Cancer (PCa) represent the most lethal and common urological cancers, respectively [1].

Kidney cancer represents 403,000 new cases and 175,000 deaths worldwide, with RCC accounting for 90% of these cases [1]. Because of kidney's anatomic location, these tumors only become symptomatic in the late stages of the disease. Even though about 60% of RCCs are incidentally detected in an early stage because of routine imaging, about 30% are still diagnosed at the symptomatic phase, which is usually associated with worse prognosis [3]. Additionally, most of the patients continue to be diagnosed with locally advanced disease, with about 17% of them presenting distant metastasis at the diagnosis [4]. Apart from these, approximately 40% of patients submitted to surgery with curative intent will also relapse in a 5-year period [5]. Because of its radio and chemo-resistance, targeted therapies are the only agents available to manage metastatic patients, but one fourth of the patients never respond to them, and the ones who do, typically develop resistance in a median of 5–11 months of treatment [6].

On the other hand, with a world estimate of 1.2 million cases and more than 350 thousand deaths in 2018, PCa is the second most frequent cancer in men and the fifth cause of death [1]. Its treatment depends on the grade, stage and age of the patients, being the androgen deprivation therapy (ADT) one of the main therapy options because of its high dependence on the androgen pathway [7]. Despite the initial high response rates, nearly all men that undergo ADT develop resistance within 2 to 3 years, progressing to Castration Resistant PCa (CRPC) [8]. In the last few years new drugs came up as alternatives to these patients, but they present limited time benefits and patients eventually relapse [9].

The late diagnosis, the lack of accurate prognosis and disease follow up biomarkers, as well as the resistance to the existing therapies are some of the major current challenges in both prostate and renal cell carcinoma [10, 11]. Thus, there is an urgent need of more accurate and sensitive biomarkers as well as alternative therapeutic approaches in these tumor models.

Almost 10 years ago, in 2011, the reprogramming of energy metabolism was considered a hallmark of cancer and in the last few years the scientific community has devoted their time to better understand it in order to develop new therapeutic approaches and biomarkers [2]. Oxidoreductases (enzymes that catalyze electrons transfer from one molecule to another) play an important part in these deregulations since they are present in the different pathways involved in cells metabolism, namely in glucose's metabolism [12].

Glucose, as one of the major "fuels" of any cell, has its metabolism altered in most tumor models [13]. However, because cancer is a heterogeneous disease, this deregulation depends on the type of cells that the tumor arises from, being RCC and PCa a good example of such differences.

2. Glucose metabolism in renal cell carcinoma

RCC is a heterogeneous group of cancers with different genetic and molecular alterations, and histological and clinical characteristics [14]. Clear cell RCC (ccRCC) accounts for about 80% of RCC cases and the most common genetic event involved in its beginning is the copy number deletion, inactivating mutation and/or epigenetic silencing of *von Hippel–Lindau* (*VHL*) [3]. Its loss or inactivation leads to an increase of Hypoxia Inducible Factor alpha (HIF-α), triggering a hypoxic response, even in normoxic conditions, from the cell and a consequent induction of its target genes transcription [15]. These genes are

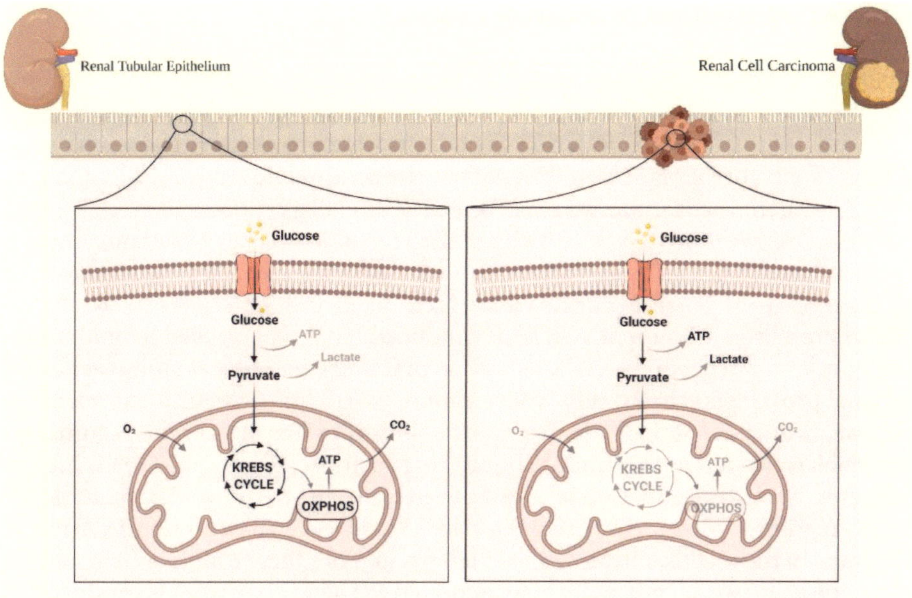

Figure 1.
RCC's glucose metabolic switch. In RCC cells, pyruvate is transformed in lactate, with the production of ATP instead of undergoing Krebs cycle and oxidative phosphorylation (Warburg effect). Created by BioRender.com.

involved in several cellular processes including glucose metabolism (*GLUT-1* and *GLUT-4*) and pH regulation (*CAIX*). Thus, ccRCC is from a very early beginning in a state of constant pseudo hypoxia [16].

This is the most likely cause of the well-known Warburg Effect which is widely documented in ccRCC [17, 18]. The Warburg Effect, or aerobic glycolysis, was firstly described in 1920 by Otto Warburg and describes cancer cells' preference to metabolize glucose through glycolysis followed by lactate fermentation instead of oxidative phosphorylation, even in the presence of oxygen (**Figure 1**) [19]. Very common in many tumors, there are several possible explanations to why cancer cells undergo these alterations, even though the energy resulting from it is significantly lower when compared to oxidative phosphorylation. Using aerobic glycolysis, cancer cells are able to obtain ATP in a faster way and this pathway supports better their high biosynthetic needs [18]. Moreover, the consequent acidification of the microenvironment due to the lactate fermentation is of great advantage to cancer cells since it has been shown to boost their invasiveness and metastatic capacity as well as to inhibit immune rejection [20, 21].

In ccRCC, besides *VHL* loss, HIF- α can also be stabilized by mechanisms like RAS activation or accumulation of Krebs cycle substrates [22]. Moreover, this effect can also be driven by the interruption of the Krebs Cycle and mutations in genes that encode enzymes like Fumarate Hydratase or Succinate Dehydrogenase, increased levels of reactive oxygen species and activation of pathways such as NRF2/KEAP1 and PI3K/mTOR [18].

In addition to thar, the deregulation of the expression of several enzymes involved in the glucose metabolic pathways has already been reported in ccRCC, including several oxidoreductases, such as glyceraldehyde-3-phosphate dehydrogenase (G3PD), lactate dehydrogenase (LDHA) which belong to the glycolysis pathway; pyruvate dehydrogenase (PD) and isocitrate dehydrogenase (IDH) involved in the Krebs Cycle and succinate dehydrogenase (SDH) which is part of the oxidative phosphorylation pathway [16, 23–26].

3. Glucose metabolism in prostate cancer

Due to its organ's function, prostatic tissue shows a unique metabolic activity under normal conditions, which will reflect in the disruptions presented by its cancer cells. One of the key functions of the prostate gland is to produce large amounts of citrate that is secreted as part of the seminal liquid [27]. Thus, normal prostate epithelial cells undergo a very inefficient energy metabolism.

In most organs, glucose is metabolized through glycolysis in pyruvate, which is decarboxylated in the mitochondria to generate Acetyl-CoA. This metabolite reacts with oxalocetate to generate citrate which is oxidized and undergoes the Krebs Cycle where a large amount of NADH is produced (that will be used in oxidative phosphorylation to produce ATP), as well as precursors of several amino acids [28]. In normal prostate epithelial cells, there is an impairment of the mitochondrial aconitase, responsible for citrate oxidization, granting this metabolite accumulation, which is needed in the seminal liquid composition [27]. Aconitase's inhibition is triggered by an accumulation of zinc in these cells due to the overexpression of the zinc-regulated transporter/iron-regulated transporter-like protein 1 (ZIP1) [29]. Thus, in these cells, citrate is the final product of glucose metabolism and oxaloacetic acid (which normally is regenerated in the Krebs cycle) is produced through aspartate imported from the plasma through a specific carrier [30]. Because of Krebs cycle inhibition, and consequent oxidative phosphorylation impairment, these cells show a higher glycolytic rate [28].

Prostate cancer cells, however, have increased energy demands. Franklin and Costello have concluded that an early event in PCa carcinogenesis is the completion of the Krebs cycle and subsequent ability to produce much more ATP [31]. In fact, PCa cells show dramatically reduced levels of zinc, and consequent reactivation of m-aconitase and of Krebs cycle [32]. Interestingly, zinc has also been shown to induce apoptosis and inhibit invasion and angiogenesis in PCa cells [33, 34].

Nevertheless, it is important to take into consideration that cells need to readjust their bioenergetics and metabolism according to their energetic needs, during cancer progression. Thus, in its metastatic stage, PCa has been shown to switch to Warburg Effect [35]. The exact mechanisms behind this switch are not yet fully understood, but the microenvironment in the metastatic sites seems to play a key

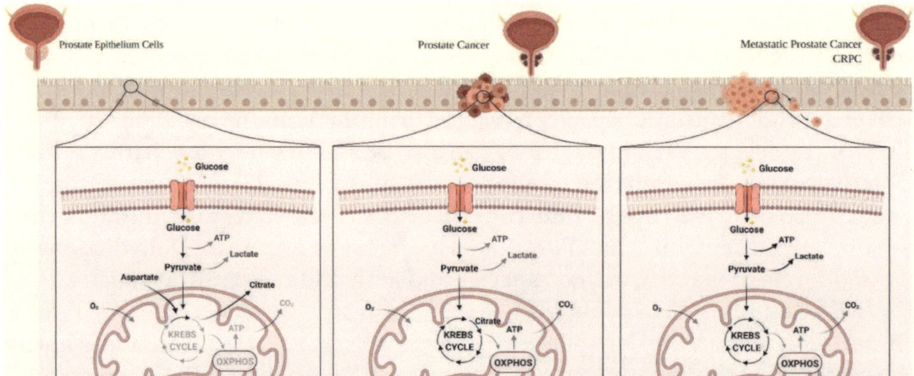

Figure 2.
PCa's glucose metabolic switch. Normal prostate epithelial cells have the Krebs cycle interrupted because of their need to secrete citrate as part of seminal fluid. In prostate cancer, Krebs cycle is resumed because of the increased demand for energy. Warburg effect is only observed in the more advanced stages of the tumor. Created by BioRender.com.

role, whether through the neighboring adipocytes or through the immune system. These seem to be able to increase HIF1α's production inducing aerobic glycolysis and blocking oxidative phosphorylation (**Figure 2**) [36, 37].

Several oxidoreductases involved in the glucose metabolic pathways have already been studied in PCa and reported as deregulated, such as glyceraldehyde-3-phosphate dehydrogenase (G3PD) and lactate dehydrogenase (LDHA) which belong to the glycolysis pathway and pyruvate dehydrogenase (PD) and isocitrate dehydrogenase (IDH) involved in the Krebs Cycle [38–41].

4. miRNAs as glucose metabolism regulators

The deregulation of the oxidoreductases as well as other enzymes involved in the glucose metabolism pathways is necessary for its reprogramming. This deregulation has already been connected with microRNAs (miRNAs), both in RCC and in PCa [18, 42].

miRNAs are short non-coding RNAs (~19 to 25 nucleotides) which regulate gene expression at a post-transcriptional level. Through binding to the 3′ untranslated region (3' UTR) of mRNAs, miRNAs induce their degradation or translation repression [43]. These molecules are important modulators of cellular behavior being involved in different biological processes such as cell development, differentiation, apoptosis, proliferation, and metabolism. This is due to their dynamic expression since each miRNA regulates up to 100 different mRNAs and more than 10,000 mRNAs are regulated by miRNAs [44].

There are different characteristics that make miRNAs good biomarkers' candidates. Firstly, miRNAs have different expression patterns in normal cells when compared with tumoral ones, and even among different subtypes or in different stages of the disease, which shows their potential as biomarkers' candidates [45]. Secondly, there has been cumulating evidence regarding the fact that miRNAs are secreted into several body fluids, such as serum, plasma, saliva or urine [46]. Finally, miRNAs circulate in these fluids incorporated into protein complexes or extracellular vesicles, which protect them from RNAse degradation and make them resistant to extreme conditions like temperature or pH differences [47].

In fact, in previous studies circulating miRNAs profiles have already been associated with histology, staging and clinical endpoints, including patients' survival and therapy response both in ccRCC and in PCa [48–50].

Thus, the study of miRNAs whose targets are involved in the glucose metabolism in tumor models such as RCC and PCa is highly important, not only because it can help to better understand the differences in metabolic deregulations of the different tumors, but also because this knowledge can be applied in designing new-targeted therapies and biomarkers.

5. Literature review and data collection

This chapter is focused on the three main glycolytic pathways: glycolysis, Krebs cycle and Lactate Fermentation. Since oxidoreductases are present in these three pathways, we chose this type of enzymes to select miRNAs that directly regulate them (**Table 1**).

Following, we used miRTarBase (version 8.0), the largest known online database of validated miRNA:mRNA target interactions, to select the miRNAs that directly target these enzymes [51]. Only studies featuring hsa-miRNAs and functional miRNA Target Interaction (MTI) evidence were considered. The selected miRNAs and the respective validated targets are displayed in **Figure 3**.

Glycolysis	Lactate fermentation	Krebs cycle
GAPDH	LDHA	PDH
		IDH
		KGDH
		SDH
		MDH

Table 1.
Oxidoreductases in glycolysis, lactate fermentation and Krebs cycle.

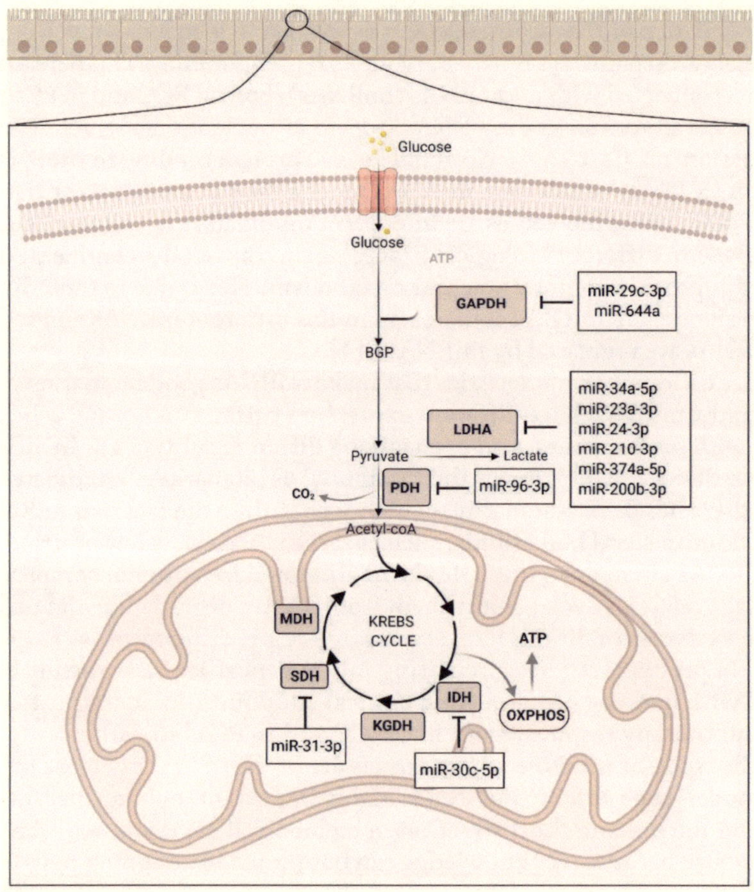

Figure 3.
miRNAs that directly regulate the oxidoreductases involved in the main pathways of glucose catabolism. Created by BioRender.com.

A systematic search in Pubmed was then conducted regarding the existing evidence for each miRNA in both ccRCC and in PCa, in order to get a deeper knowledge of these miRNAs expression in these tumor models. To do so, we combined each miRNA with the following keywords: "renal cell carcinoma", "RCC", "Kidney Cancer"; "Prostate Cancer". The obtained scientific papers were manually curated to determine the association between the miRNA and either RCC or PCa. The criteria of exclusion were the following: 1) scientific papers that do not report results from human samples; 2) scientific papers that do not directly correlate the miRNA with the disease. From the 56 papers initially found, 23 were excluded. For each selected paper, we extracted information regarding the deregulation of the miRNA's expression in each tumor model (upregulated ↑/downregulated ↓) and gathered it in the following tables, according to the metabolic processes involved.

5.1 Glycolysis

Glycolysis is the pathway responsible for converting glucose in pyruvate and it is constituted by a series of enzymatic reactions. Its sixth step is catalyzed by an oxido-reductase - Glyceraldehyde_3-phosphate_dehydrogenase (GAPDH) – responsible for transforming glyceraldehyde 3-phosphate in D-glycerate 1,3-biphosphate. According to miRTarBase (version 8.0), GAPDH is directly targeted by miR-29c-3p and miR-644a [51]. The studies regarding these miRNAs in both RCC and PCa are scarce, and miR-644a's expression is still not described in RCC nor miR-29c-3p's expression is described in PCa (**Tables 2** and **3**).

In both tumor models, the miRNAs targeting GAPDH are downregulated, which may partly explain the upregulation of GADPH already observed in PCa [52, 53]. There is, in fact, an increase of glucose consumption in cancer due to the bigger energetic needs of tumoral cells. Since glycolysis is the basis of glucose catabolism, either by following Krebs Cycle or Lactate Fermentation, the upregulation of the expression of this pathway's enzymes will help ensure cancer cells' catabolic demands.

5.2 Lactate fermentation

Lactate fermentation is the metabolic process in which the pyruvate resulting from glycolysis is transformed in lactate with ATP production. This reaction is catalyzed by an oxidoreductase – Lactate Dehydrogenase (LDHA), whose mRNA, according to miRTarBase (version 8.0), is directly targeted by miR-34a-5p, miR-23a-3p, miR-24-3p, miR-210-3p, miR-374a-5p and miR-200b-3p [51]. To the best of our knowledge, there are still no studies regarding miR-24-3p and miR-374a-5p's expression in RCC as well as miR-374a-5p's expression in PCa. The studies regarding the other miRNA's expression in RCC are summarized in **Table 4** and the ones regarding miRNA's expression in PCa are summarized in **Table 5**.

In RCC, the available studies for the selected miRNAs are controversial. This may be result of lack of standardized procedures but also of the different types of samples analyzed. Moreover, it is interesting to look at the studies of miR-210-3p's expression. This miRNA was significantly increased in ccRCC patients at the time of surgery, when compared to healthy donors, but significantly decreased in follow-up disease-free ccRCC patients of the same cohort [62, 64]. These studies show, not only this miRNA potential as follow-up biomarker but are also an example

Enzyme	miRNA	Sample type	Outcome	References
GAPDH	miR-29c-3p	Tissues and Cell lines	↓	[52]

Table 2.
Deregulation of the miRNAs that directly target the glycolysis' oxidoreductases in RCC.

Enzyme	miRNA	Sample type	Outcome	References
GAPDH	miR-644a	Tissues	↓	[53]

Table 3.
Deregulation of the miRNAs that directly target the glycolysis' oxidoreductases in PCa.

Enzyme	miRNA	Sample type	Outcome	References
LDHA	miR-34a-5p	Cell lines	↑	[54]
		Tissues and cell lines	↑	[55]
		Tissues	↓	[56]
	miR-23a-3p	Cell lines	↓	[57]
		Tissues and cell lines	↑	[58]
	miR-210-3p	Tissues	↑	[59]
		Cell lines	↓	[60]
		Cell lines	↓	[61]
		Tissues and urine	↑↓	[62]
		Tissues	↑	[63]
		Tissues and urine	↑↓	[64]
		Tissues	↑	[65]
		Cell lines and plasma	↑	[66]
	miR-200b-3p	Cell lines	↓	[67]

Table 4.
Deregulation of the miRNAs that directly target the lactate Fermentation's oxidoreductases in RCC.

of miRNAs dynamic expression. In PCa, one can notice that hormonal resistant and metastatic PCa show a decrease in miR-34a-5p and miR-200b-3p, which may traduce in an increase of LDHA and the switch to Warburg Effect which is only observed in these stages of PCa [68, 79].

5.3 Krebs cycle

Krebs Cycle, also known as the tricarboxylic acid cycle, follows glycolysis in the glucose catabolism when oxygen is present. It is preceded by the transformation of pyruvate in acetyl-coA, which will enter the cycle – a series of reactions that provide precursors of amino acids as well as the reducing agent NADH which will be used in the oxidative phosphorylation pathway and lead to ATP production.

Pyruvate oxidation in acetyl-coA is catalyzed by an oxidoreductase – Pyruvate dehydrogenase (PDH), whose mRNA is, according to miRTarBase (version 8.0) directly targeted by miR-96-3p [51]. However, there are still no studies regarding this miRNA in both RCC and in PCa.

In the series of reactions of Krebs Cycle, there are 4 reactions catalyzed by 4 oxidoreductases – Isocitrate dehydrogenase (IDH), α-ketoglutarate dehydrogenase (KDGH), Succinate dehydrogenase (SDH) and Malate dehydrogenase (MDH). According to miRTarBase (version 8.0), miRNAs directly targeting KDGH and MDH were not yet identified. Moreover, SDH is directly targeted by miR-31-3p, which, to the best of our knowledge, has not yet been studied in RCC and in PCa [51].

IDH is directly targeted by miR-30c-5p. There are few studies regarding this miRNA both in RCC (**Table 6**) and in PCa (**Table 7**).

In these studies, the expression of miR-30c-5p in RCC is downregulated which would suggest an upregulation of IDH's mRNA expression. However, this protein was shown to be downregulated in this tumor model [85]. In fact, a single mRNA can be regulate by several miRNAs, making the miRNA:mRNA expression not always inversely correlated. Nevertheless, the fact that miR-30-c-5p was

Enzyme	miRNA	Sample type	Outcome	References
LDHA	miR-34a-5p	Cell lines (resistant vs. hormonal sensitive)	↓	[68]
		Urinary exosomes and tissues	↓	[69]
		Cell lines	↓	[70]
	miR-23a-3p	Tissues	↑	[71]
	miR-24-3p	Urine	↓	[72]
		Urine	↓	[73]
		Tissues and cell lines	↓	[74]
	miR-210-3p	Tissues	↑	[75]
		Tissues	↑	[76]
	miR-200b-3p	Tissues	↑	[77]
		Cell lines	↓	[78]
		Metastatic tissues	↓	[79]
		Chemo-resistant cells	↑	[80]

Table 5.
Deregulation of the miRNAs that directly target the lactate fermentation's oxidoreductases in PCa.

Enzyme	miRNA	Sample type	Outcome	References
IDH	miR-30c-5p	Urinary exosomes	↓	[81]
		Tissues	↓	[82]

Table 6.
Deregulation of the miRNAs that directly target the Krebs cycle's oxidoreductases in RCC.

Enzyme	miRNA	Sample type	Outcome	References
IDH	miR-30c-5p	Tissues	↓	[83]
		Urine	↑	[73]
		Tissues	↑	[84]

Table 7.
Deregulation of the miRNAs that directly target the Krebs Cycle's oxidoreductases in PCa.

downregulated in urinary exosomes shows its potential as a biomarker in a liquid biopsy approach [81].

In PCa the results regarding this miRNA are scarce and controversial, showing the need of more studies to clarify its expression levels.

6. Discussion

miRNAs potential in the oncology field has been widely recognized and there has been an increase of studies regarding their deregulation in cancer in the last few years. However, there are many genes whose mRNA have not been identified as direct targets of any miRNA. In this book chapter, both KGDH and MDH, key enzymes in the Krebs Cycle, have not been directly associated with any miRNAs. Moreover, there are several miRNAs that directly target the mRNA of key enzymes of glucose catabolism but have not yet been studied in RCC (miR-644a, miR-24-3p, miR-374a-5p, miR-96-3p and miR-31-3p) and in PCa (miR-29c-3p, miR-374a-5p, miR96-3p, miR-31-3p). Additionally, some miRNAs present controversial results which shall be subject of more studies to confirm their deregulation. Nevertheless, two miRNAs have been identified as downregulated (miR-29c-3p and miR-200b-2p) in RCC and three miRNAs have been identified as downregulated (miR-644a, miR-34a-5p and miR-24-3p) and two as upregulated (miR-23a-3p and miR-210-3p) in PCa. Their potential as biomarkers of both RCC and PCa could be increased if combined as a profile, which could pose as an advance to establish a successful liquid biopsy approach.

Because of their influence in their target genes' expression, the reestablishment of miRNAs' levels may have a great impact in the regulation of glucose metabolism. Restoring the levels of the downregulated miRNAs in both RCC and PCa could benefit the current cancer therapies and one possible way to do so is through a nanomedicine approach. Nanoparticles (NPs) are small organized structures with sizes between in size 1 and 100 nm that show very specific chemical and physical properties due to their size and composition [86]. Even though the existing research is scarce, NPs can improve the specificity of miRNAs delivery to target cells (thus reducing side effects) and allow for controlled miRNA release [87]. They also can protect them from degradation and prevent their clearance by the reticuloendothelial system. Moreover, they avoid unfavorable immune cell stimulation [87]. NPs highly depend on their capping which acts prevents their agglomeration and stops uncontrolled growth. The choice of capping will highly influence NPs properties. To effectively deliver the miRNAs selected in this chapter, a glucose capping could be an interesting choice. As stated above, both in RCC and PCa, tumoral cells show an increased glucose consumption when compared with their counterpart normal cells. Thus, glucose as NP's capping could favor the selective delivery of miRNAs and would likely not be recognized as antagonist by the immune system.

7. Conclusions

The deregulation of glucose metabolism as a great influence in the pathophysiology of cancer, with the oxidoreductases involved in its pathways posing as both an opportunity to better comprehend the disease and finding not only strategies of detecting and monitoring it but also new therapeutic strategies. miRNAs could be part of these strategies since they influence the expression of these enzymes. Both in RCC and PCa, there are studies regarding miRNAs that target these oxidoreductases, showing their impact in patients' prognosis. In the future, more studies are needed, regarding the identification of more miRNAs that target for example KGDH and MDH and their validation in RCC and PCa. Moreover, exploring the potential of glucose capped NPs carrying these miRNAs could help establish new therapeutic strategies that would benefit RCC and PCa management.

Acknowledgements

We would like to thank Liga Portuguesa Contra o Cancro – Núcleo Regional do Norte – LPCC-NRN (Portuguese League against Cancer). M.M. is a recipient of a research scholarship awarded by LPCC-NRN (Portuguese League against Cancer—Northern Branch). F.D. is a recipient of a research fellowship from the project NORTE-01-0247-FEDER-033399, supported by Norte Portugal regional Operational Program (NORTE 2020), under the PORTUGAL 2020 Partnership Agreement, through European Regional Development Fund (ERDF).

Conflict of interest

The authors declare no conflict of interest.

Author details

Mariana Gomes Morais[1,2,3], Francisca Guilherme Carvalho Dias[1], João Alexandre Velho Prior[4], Ana Luísa Pereira Teixeira[1]* and Rui Manuel de Medeiros Melo Silva[1,2,3,5,6]

1 Molecular Oncology and Viral Pathology Group, IPO-Porto Research Center (CI-IPOP), Portuguese Oncology Institute of Porto (IPO-Porto), Research Center- LAB2, E Bdg 1st floor, Rua Dr António Bernardino de Almeida, Porto, Portugal

2 Institute of Biomedical Sciences Abel Salazar, University of Porto (ICBAS-UP), Porto, Portugal

3 Research Department of the Portuguese League Against Cancer Regional Nucleus of the North (LPCC-NRN), Porto, Portugal

4 LAQV, REQUIMTE, Laboratory of Applied Chemistry, Department of Chemical Sciences, Faculty of Pharmacy, University of Porto, Porto, Portugal

5 Faculty of Medicine, University of Porto (FMUP), Alameda Prof. Hernâni Monteiro, Porto, Portugal

6 Biomedical Research Center (CEBIMED), Faculty of Health Sciences of Fernando Pessoa University (UFP), Porto, Portugal

*Address all correspondence to: ana.luisa.teixeira@ipoporto.min-saude.pt

References

[1] Bray, F., et al., *Global cancer statistics 2018: GLOBOCAN estimates of incidence and mortality worldwide for 36 cancers in 185 countries.* CA: A Cancer Journal for Clinicians, 2018. **68**(6): p. 394-424.

[2] Hanahan, D. and R.A. Weinberg, *Hallmarks of cancer: the next generation.* Cell, 2011. **144**(5): p. 646-74.

[3] Escudier, B., et al., *Renal cell carcinoma: ESMO Clinical Practice Guidelines for diagnosis, treatment and follow-up†.* Ann Oncol, 2019. **30**(5): p. 706-720.

[4] Capitanio, U. and F. Montorsi, *Renal cancer.* The Lancet. **387**(10021): p. 894-906.

[5] Dabestani, S., et al., *Renal cell carcinoma recurrences and metastases in primary non-metastatic patients: a population-based study.* World J Urol, 2016. **34**(8): p. 1081-6.

[6] Ravaud, A. and M. Gross-Goupil, *Overcoming resistance to tyrosine kinase inhibitors in renal cell carcinoma.* Cancer Treat Rev, 2012. **38**(8): p. 996-1003.

[7] Wang, G., et al., *Genetics and biology of prostate cancer.* Genes Dev, 2018. **32**(17-18): p. 1105-1140.

[8] Tucci, M., et al., *Enzalutamide-resistant castration-resistant prostate cancer: challenges and solutions.* Onco Targets Ther, 2018. **11**: p. 7353-7368.

[9] Antonarakis, E.S., *Current understanding of resistance to abiraterone and enzalutamide in advanced prostate cancer.* Clin Adv Hematol Oncol, 2016. **14**(5): p. 316-9.

[10] Hsieh, J.J., et al., *Renal cell carcinoma.* Nat Rev Dis Primers, 2017. **3**: p. 17009.

[11] Dong, L., et al., *Metastatic prostate cancer remains incurable, why?* Asian J Urol, 2019. **6**(1): p. 26-41.

[12] Sellés Vidal, L., et al., *Review of NAD(P)H-dependent oxidoreductases: Properties, engineering and application.* Biochimica et Biophysica Acta (BBA) - Proteins and Proteomics, 2018. **1866**(2): p. 327-347.

[13] Lin, X., et al., *Glucose Metabolism on Tumor Plasticity, Diagnosis, and Treatment.* Front Oncol, 2020. **10**: p. 317.

[14] Moch, H., et al., *The 2016 WHO Classification of Tumours of the Urinary System and Male Genital Organs—Part A: Renal, Penile, and Testicular Tumours.* European Urology. **70**(1): p. 93-105.

[15] Baldewijns, M.M., et al., *VHL and HIF signalling in renal cell carcinogenesis.* J Pathol, 2010. **221**(2): p. 125-38.

[16] Schodel, J., et al., *Hypoxia, Hypoxia-inducible Transcription Factors, and Renal Cancer.* Eur Urol, 2016. **69**(4): p. 646-57.

[17] Weiss, R.H., *Metabolomics and Metabolic Reprogramming in Kidney Cancer.* Semin Nephrol, 2018. **38**(2): p. 175-182.

[18] Morais, M., et al., *MicroRNAs and altered metabolism of clear cell renal cell carcinoma: Potential role as aerobic glycolysis biomarkers.* Biochim Biophys Acta, 2017. **1861**(9): p. 2175-2185.

[19] Koppenol, W.H., P.L. Bounds, and C.V. Dang, *Otto Warburg's contributions to current concepts of cancer metabolism.* Nat Rev Cancer, 2011. **11**(5): p. 325-37.

[20] Boedtkjer, E. and S.F. Pedersen, *The Acidic Tumor Microenvironment as a Driver of Cancer.* Annu Rev Physiol, 2020. **82**: p. 103-126.

[21] McCarty, M.F. and J. Whitaker, *Manipulating tumor acidification as a cancer treatment strategy.* Altern Med Rev, 2010. **15**(3): p. 264-72.

[22] Wigerup, C., S. Påhlman, and D. Bexell, *Therapeutic targeting of hypoxia and hypoxia-inducible factors in cancer.* Pharmacology & Therapeutics, 2016. **164**: p. 152-169.

[23] Kim, J.-w., et al., *HIF-1-mediated expression of pyruvate dehydrogenase kinase: A metabolic switch required for cellular adaptation to hypoxia.* Cell Metabolism, 2006. **3**(3): p. 177-185.

[24] Kinnaird, A., et al., *Metabolic Modulation of Clear-cell Renal Cell Carcinoma with Dichloroacetate, an Inhibitor of Pyruvate Dehydrogenase Kinase.* European Urology, 2016. **69**(4): p. 734-744.

[25] Vilà, M.R., et al., *Increased glyceraldehyde-3-phosphate dehydrogenase expression in renal cell carcinoma identified by RNA-based, arbitrarily primed polymerase chain reaction.* Cancer, 2000. **89**(1): p. 152-164.

[26] Tsai, T.-H. and W.-Y. Lee, *Succinate Dehydrogenase–Deficient Renal Cell Carcinoma.* Archives of Pathology & Laboratory Medicine, 2018. **143**(5): p. 643-647.

[27] Eidelman, E., et al., *The Metabolic Phenotype of Prostate Cancer.* Front Oncol, 2017. **7**: p. 131.

[28] Cutruzzolà, F., et al., *Glucose Metabolism in the Progression of Prostate Cancer.* Front Physiol, 2017. **8**: p. 97.

[29] Costello, L.C., et al., *Human prostate cancer ZIP1/zinc/citrate genetic/metabolic relationship in the TRAMP prostate cancer animal model.* Cancer Biol Ther, 2011. **12**(12): p. 1078-84.

[30] Franklin, R.B., et al., *EAAC1 is expressed in rat and human prostate epithelial cells; functions as a high-affinity L-aspartate transporter; and is regulated by prolactin and testosterone.* BMC Biochem, 2006. **7**: p. 10.

[31] Costello, L.C. and R.B. Franklin, *The intermediary metabolism of the prostate: a key to understanding the pathogenesis and progression of prostate malignancy.* Oncology, 2000. **59**(4): p. 269-82.

[32] Costello, L.C. and R.B. Franklin, *Decreased zinc in the development and progression of malignancy: an important common relationship and potential for prevention and treatment of carcinomas.* Expert Opinion on Therapeutic Targets, 2017. **21**(1): p. 51-66.

[33] Uzzo, R.G., et al., *Zinc inhibits nuclear factor-kappa B activation and sensitizes prostate cancer cells to cytotoxic agents.* Clin Cancer Res, 2002. **8**(11): p. 3579-83.

[34] Feng, P., et al., *Direct effect of zinc on mitochondrial apoptogenesis in prostate cells.* Prostate, 2002. **52**(4): p. 311-8.

[35] Gonzalez-Menendez, P., et al., *The dark side of glucose transporters in prostate cancer: Are they a new feature to characterize carcinomas?* Int J Cancer, 2018. **142**(12): p. 2414-2424.

[36] Diedrich, J.D., et al., *Bone marrow adipocytes promote the Warburg phenotype in metastatic prostate tumors via HIF-1α activation.* Oncotarget, 2016. **7**(40): p. 64854-64877.

[37] Vaughan, R.A., et al., *Tumor necrosis factor alpha increases aerobic glycolysis and reduces oxidative metabolism in prostate epithelial cells.* Prostate, 2013. **73**(14): p. 1538-46.

[38] Harada, N., et al., *Glyceraldehyde-3-phosphate dehydrogenase enhances transcriptional activity of androgen receptor in prostate cancer cells.* J Biol Chem, 2007. **282**(31): p. 22651-61.

[39] Li, F., et al., *Association between lactate dehydrogenase levels and oncologic outcomes in metastatic prostate cancer: A meta-analysis.* Cancer Medicine. **n/a**(n/a).

[40] Pereira-Nunes, A., et al., *Targeting lactate production and efflux in prostate cancer.* Biochimica et Biophysica Acta (BBA) - Molecular Basis of Disease, 2020: p. 165894.

[41] Gonthier, K., et al., *Reprogramming of Isocitrate Dehydrogenases Expression and Activity by the Androgen Receptor in Prostate Cancer.* Mol Cancer Res, 2019. **17**(8): p. 1699-1709.

[42] Kasomva, K., et al., *Roles of microRNA in prostate cancer cell metabolism.* Int J Biochem Cell Biol, 2018. **102**: p. 109-116.

[43] Lin, S. and R.I. Gregory, *MicroRNA biogenesis pathways in cancer.* Nat Rev Cancer, 2015. **15**(6): p. 321-33.

[44] Hummel, R., D.J. Hussey, and J. Haier, *MicroRNAs: predictors and modifiers of chemo- and radiotherapy in different tumour types.* Eur J Cancer, 2010. **46**(2): p. 298-311.

[45] Wang, H., et al., *Circulating microRNAs as potential cancer biomarkers: the advantage and disadvantage.* Clin Epigenetics, 2018. **10**: p. 59.

[46] Weber, J.A., et al., *The microRNA spectrum in 12 body fluids.* Clin Chem, 2010. **56**(11): p. 1733-41.

[47] Duttagupta, R., et al., *Impact of Cellular miRNAs on Circulating miRNA Biomarker Signatures.* PLOS ONE, 2011. **6**(6): p. e20769.

[48] Nogueira, I., et al., *Everolimus resistance in clear cell renal cell carcinoma: miRNA-101 and HIF-2alpha as molecular triggers?* Future Oncol, 2019. **15**(20): p. 2361-2370.

[49] Dias, F., et al., *Extracellular Vesicles Enriched in hsa-miR-301a-3p and hsa-miR-1293 Dynamics in Clear Cell Renal Cell Carcinoma Patients: Potential Biomarkers of Metastatic Disease.* Cancers (Basel), 2020. **12**(6).

[50] Santos, J.I., et al., *Influence of peripheral whole-blood microRNA-7 and microRNA-221 high expression levels on the acquisition of castration-resistant prostate cancer: evidences from in vitro and in vivo studies.* Tumour Biol, 2014. **35**(7): p. 7105-13.

[51] Huang, H.-Y., et al., *miRTarBase 2020: updates to the experimentally validated microRNA–target interaction database.* Nucleic Acids Research, 2019. **48**(D1): p. D148-D154.

[52] Chen, J., et al., *Overexpressed pseudogenes, DUXAP8 and DUXAP9, promote growth of renal cell carcinoma and serve as unfavorable prognostic biomarkers.* Aging, 2019. **11**(15): p. 5666-5688.

[53] Ebron, J.S., et al., *MiR-644a Disrupts Oncogenic Transformation and Warburg Effect by Direct Modulation of Multiple Genes of Tumor-Promoting Pathways.* Cancer Research, 2019. **79**(8): p. 1844-1856.

[54] Wang, K., et al., *Androgen receptor regulates ASS1P3/miR-34a-5p/ASS1 signaling to promote renal cell carcinoma cell growth.* Cell Death & Disease, 2019. **10**(5): p. 339.

[55] Jing, Z.-F., et al., *Inhibition of miR-34a-5p can rescue disruption of the p53-DAPK axis to suppress progression of clear cell renal cell carcinoma.* Molecular Oncology, 2019. **13**(10): p. 2079-2097.

[56] Jin, C., et al., *Circ_0039569 promotes renal cell carcinoma growth and metastasis by regulating miR-34a-5p/CCL22.* Am J Transl Res, 2019. **11**(8): p. 4935-4945.

[57] Ishihara, T., et al., *Expression of the Tumor Suppressive miRNA-23b/27b Cluster is a Good Prognostic Marker in Clear Cell Renal Cell Carcinoma.* Journal of Urology, 2014. **192**(6): p. 1822-1830.

[58] Quan, J., et al., *MiR-23a-3p acts as an oncogene and potential prognostic biomarker by targeting PNRC2 in RCC.* Biomedicine & Pharmacotherapy, 2019. **110**: p. 656-666.

[59] Zhang, J., et al., *Global and Targeted miRNA Expression Profiling in Clear Cell Renal Cell Carcinoma Tissues Potentially Links miR-155-5p and miR-210-3p to both Tumorigenesis and Recurrence.* Am J Pathol, 2018. **188**(11): p. 2487-2496.

[60] Li, S., et al., *Down-regulation of miR-210-3p encourages chemotherapy resistance of renal cell carcinoma via modulating ABCC1.* Cell & Bioscience, 2018. **8**(1): p. 9.

[61] Yoshino, H., et al., *microRNA-210-3p depletion by CRISPR/Cas9 promoted tumorigenesis through revival of TWIST1 in renal cell carcinoma.* Oncotarget, 2017. **8**(13).

[62] Petrozza, V., et al., *Secreted miR-210-3p as non-invasive biomarker in clear cell renal cell carcinoma.* Oncotarget, 2017. **8**(41).

[63] Petrozza, V., et al., *Oncogenic MicroRNAs Characterization in Clear Cell Renal Cell Carcinoma.* Int J Mol Sci, 2015. **16**(12): p. 29219-25.

[64] Petrozza, V., et al., *Emerging role of secreted miR-210-3p as potential biomarker for clear cell Renal Cell Carcinoma metastasis.* Cancer Biomarkers, 2020. **27**: p. 181-188.

[65] Shiomi, E., et al., *Analysis of Expression Patterns of MicroRNAs That Are Closely Associated With Renal Carcinogenesis.* Frontiers in Oncology, 2019. **9**(431).

[66] Dias, F., et al., *Plasmatic miR-210, miR-221 and miR-1233 profile: potential liquid biopsies candidates for renal cell carcinoma.* Oncotarget, 2017. **8**(61): p. 103315-103326.

[67] Duns, G., et al., *The entire miR-200 seed family is strongly deregulated in clear cell renal cell cancer compared to the proximal tubular epithelial cells of the kidney.* Genes, Chromosomes and Cancer, 2013. **52**(2): p. 165-173.

[68] Ma, Y., et al., *Long noncoding RNA DANCR contributes to docetaxel resistance in prostate cancer through targeting the miR-34a-5p/JAG1 pathway.* Onco Targets Ther, 2019. **12**: p. 5485-5497.

[69] Rodríguez, M., et al., *Identification of non-invasive miRNAs biomarkers for prostate cancer by deep sequencing analysis of urinary exosomes.* Molecular Cancer, 2017. **16**(1): p. 156.

[70] Jiang, X., et al., *LncRNA NEAT1 promotes docetaxel resistance in prostate cancer by regulating ACSL4 via sponging miR-34a-5p and miR-204-5p.* Cellular Signalling, 2020. **65**: p. 109422.

[71] Strand, S.H., et al., *Validation of the four-miRNA biomarker panel MiCaP for prediction of long-term prostate cancer outcome.* Scientific Reports, 2020. **10**(1): p. 10704.

[72] Fredsøe, J., et al., *Diagnostic and Prognostic MicroRNA Biomarkers for Prostate Cancer in Cell-free Urine.* Eur Urol Focus, 2018. **4**(6): p. 825-833.

[73] Fredsøe, J., et al., *Independent Validation of a Diagnostic Noninvasive 3-MicroRNA Ratio Model (uCaP) for Prostate Cancer in Cell-Free Urine.* Clinical Chemistry, 2019. **65**(4): p. 540-548.

[74] Li, X., et al., *Knockdown of lncRNA CCAT1 enhances sensitivity of paclitaxel in prostate cancer via regulating*

miR-24-3p and FSCN1. Cancer Biology & Therapy, 2020. **21**(5): p. 452-462.

[75] Ren, D., et al., *Oncogenic miR-210-3p promotes prostate cancer cell EMT and bone metastasis via NF-κB signaling pathway.* Molecular Cancer, 2017. **16**(1): p. 117.

[76] Dai, Y., et al., *The TGF-β signalling negative regulator PICK1 represses prostate cancer metastasis to bone.* British Journal of Cancer, 2017. **117**(5): p. 685-694.

[77] Janiak, M., et al., *TIMP4 expression is regulated by miR-200b-3p in prostate cancer cells.* APMIS, 2017. **125**(2): p. 101-105.

[78] He, M., et al., *Down-regulation of miR-200b-3p by low p73 contributes to the androgen-independence of prostate cancer cells.* The Prostate, 2013. **73**(10): p. 1048-1056.

[79] Xia, L., et al., *Transcriptional regulation of PRKAR2B by miR-200b-3p/200c-3p and XBP1 in human prostate cancer.* Biomedicine & Pharmacotherapy, 2020. **124**: p. 109863.

[80] Samli, H., et al., *Paclitaxel resistance and the role of miRNAs in prostate cancer cell lines.* World Journal of Urology, 2019. **37**(6): p. 1117-1126.

[81] Song, S., et al., *Urinary exosome miR-30c-5p as a biomarker of clear cell renal cell carcinoma that inhibits progression by targeting HSPA5.* Journal of Cellular and Molecular Medicine, 2019. **23**(10): p. 6755-6765.

[82] Onyshchenko, K.V., et al., *Expression of micro-RNA hsa-miR-30c-5p and hsa-miR-138-1 in renal cell carcinoma.* Exp Oncol, 2020. **42**(2): p. 115-119.

[83] Cochetti, G., et al., *Different levels of serum microRNAs in prostate cancer and benign prostatic hyperplasia: evaluation of potential diagnostic and prognostic role.* Onco Targets Ther, 2016. **9**: p. 7545-7553.

[84] Zhao, Z., et al., *A Novel Predictor Tool of Biochemical Recurrence after Radical Prostatectomy Based on a Five-MicroRNA Tissue Signature.* Cancers, 2019. **11**(10): p. 1603.

[85] Laba, P., J. Wang, and J. Zhang, *Low level of isocitrate dehydrogenase 1 predicts unfavorable postoperative outcomes in patients with clear cell renal cell carcinoma.* BMC Cancer, 2018. **18**(1): p. 852.

[86] Deepak, P., et al., *Chapter 15 - Chemical and green synthesis of nanoparticles and their efficacy on cancer cells*, in *Green Synthesis, Characterization and Applications of Nanoparticles*, A.K. Shukla and S. Iravani, Editors. 2019, Elsevier. p. 369-387.

[87] Lee, S.W.L., et al., *MicroRNA delivery through nanoparticles.* Journal of controlled release : official journal of the Controlled Release Society, 2019. **313**: p. 80-95.

Chapter 5

Steroidal 5α-Reductase: A Therapeutic Target for Prostate Disorders

Neelima Dhingra

Abstract

Steroidal 5α-reductase is a system of NADPH dependent enzyme that catalyzes the irreversible conversion of Δ^4–3-ketosteroid precursor (testosterone) to its corresponding 5α-reduced metabolite (dihydrotestosterone). Initial role of DHT was discovered through males pseudohermaphroditism, a genetic disorder with complete or partial 5α-reductase deficiency accompanied with features at critical juncture of fetal and postnatal development. However, excessive DHT production, has brought a revolution in revealing the etiology of complications like prostate cancer and benign prostatic hyperplasia. Over the last two decades, converging lines of evidences have highlighted the role of 5α-reductase inhibitors in the treatment of these androgen dependent disorders. Finasteride and Dutasteride, are the two clinically approved inhibitors available in the market, that helps in reducing the prostate volume by blocking the 5a-reductase enzyme.

Keywords: androgen, isozymes, prostate, genetic disorder, benign prostatic hyperplasia

1. Introduction

The prostate gland located between the bladder and the rectum is an heterogeneous organ, and wraps around the urethra. It is considered to be consisted of central, peripheral or transitional zone and composed of three different types of cells: glandular epithelial cells, smooth muscle cells and stromal cells (**Figure 1**). At the time of birth, prostate is about the size of a pea and undergoes many changes during the course of man's life. It grows only slightly until puberty, then it begins to enlarge rapidly attaining normal adult size and shape [1].

The gland generally remains stable until about the mid 40s, and in most men over the age of 60, the prostate begins to enlarge. The dense capsule surrounding the enlarging prostate prevents it from further expansion outward, which in turn forces the prostate to press against the urethra, and partially block urine flow (**Figure 2**). This apparent increase in number of stromal and epithelial cells results in obstruction of the proximal urethera and condition is called as benign prostatic hyperplasia (BPH). This obstruction, in turn causes bladder irritation and contraction, even for small amount of urine. Eventually the bladder weakens and does not completely empty through urination [2].

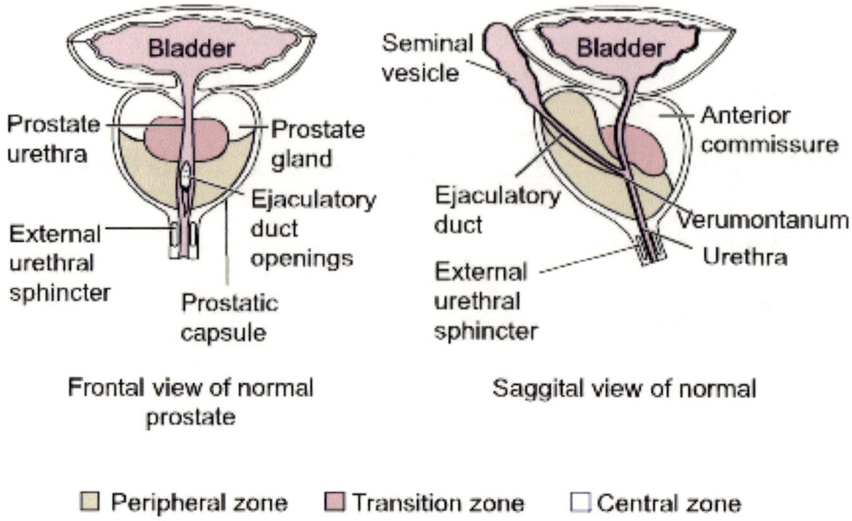

Figure 1.
Location and different sections of prostate gland.

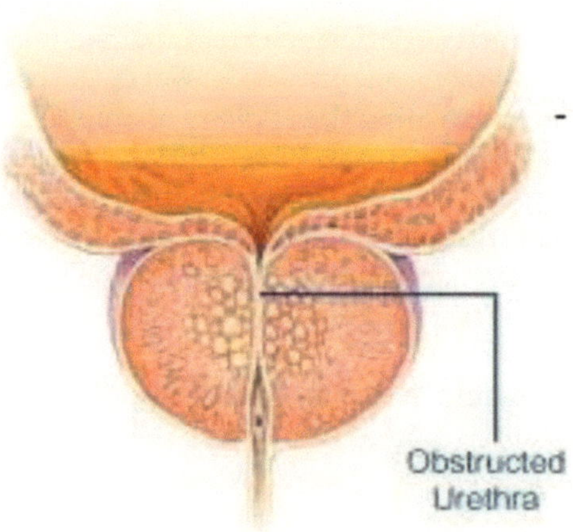

Figure 2.
Enlarged prostate gland.

Clinically BPH is manifested as lower urinary tract symptoms (LUTS) and consisting of voiding and storage symptoms such as slow urinary stream, splitting or spraying of urinary stream, recurrent urinary stream, straining to void and terminal dribbling, hesitancy, urgency, increased frequency, and incontinence. Although urge incontinence is an irritative symptom, it may indicate the presence of obstruction [3, 4].

BPH is also described as quality of life disorder, as its affects man's ability to initiate or terminate urine flow stream (the symptoms interfere with the normal activities) and reduces the feeling of well being. Though the etiology of hyerplastic process of BPH is clearly not known, but many partially overlapping and complementary theories have been proposed for the overgrowth of smooth muscle tissue and glandular epithelial tissue like aging: late activation of cell growth [5], defective cell death and hormonal changes. According to the most widely accepted hypothesis i.e. androgen (dihydrotestosterone hypothesis) BPH occurs due to an age related changes in prostate androgen metabolism that favors the accumulation of DHT and responsible for cell growth in the tissues that lines the prostate gland thus rapid prostate enlargement [6, 7].

2. Treatment options for BPH

During the last two decades, it has become clear that the management of LUTS associated with BPH is much more than just treating symptoms. It is a progressive disease and defined as worsening of symptoms, increase in prostate volume (PV), deterioration of urinary flow rate, inability to void i.e. acute urinary retention (AUR) and the need for surgery either for AUR or deteriorating symptoms [8]. Further, AUR with an annual risk of less than 1% is found to be uncommon, but requires urgent bladder catheterization. Therefore, diagnosis, monitoring, frequency, severity and assessment of the prognosis for disease progression should be assessed before management decisions. EAU guidelines have recommended a series of evaluation as a routine part of the initial assessment of men with LUTS, that includes clinical history, a validated questionnaire to assess symptoms, physical examination, creatinine measurement, urinalysis, flow rates, postvoid residual (PVR) volume and serum prostate-specific antigen (PSA) measurement especially when a diagnosis of prostatic carcinoma is required [9]. A more profound knowledge on the pathogenesis, the natural history and risk of the progression, has enabled more differentiated therapy of elderly men with lower urinary tract symptoms due to benign prostatic hyperplasia as follows (**Figure 3**) [10, 11].

2.1 Watchful waiting

Watchful waiting is a well known approach to treat BPH where men are asymptomatic or with mild to moderate symptoms without causing no serious health. It is generally considered as the first tier in the therapeutic cascade and patients are

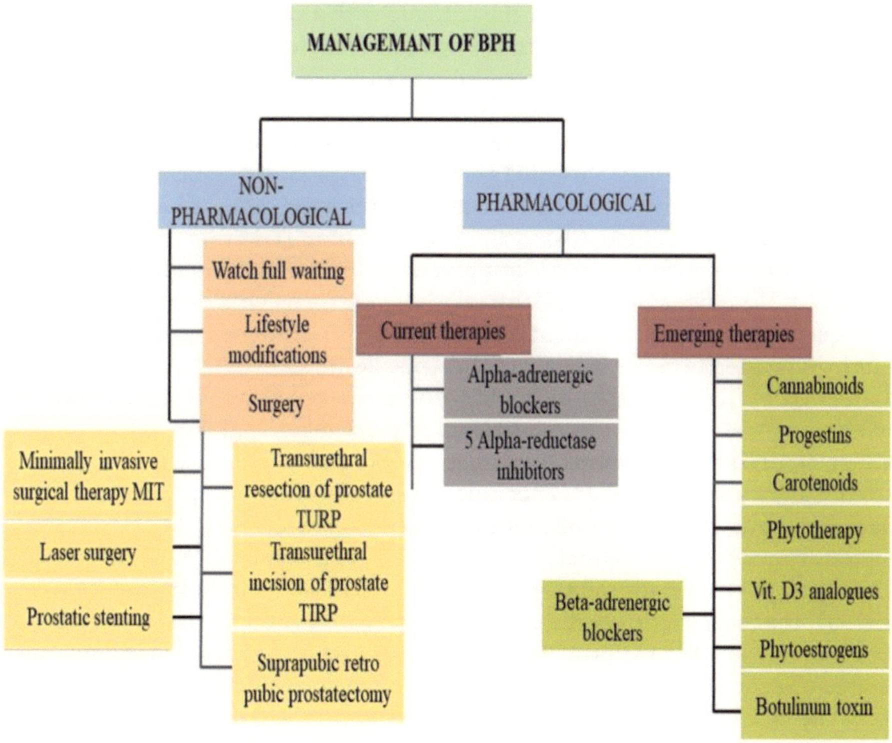

Figure 3.
Management options.

monitored by his physician without receiving any active intervention. Untreated BPH will progress to AUR and other complications such as renal insufficiency and stones. Thus regular check up along with continual education is recommended to avoid chances of occurrence of serious complications [12, 13]. Further, optimization can be achieved by including certain lifestyle or dietary changes as recommend in EAU guidelines, to prevent the deterioration requiring medical or surgical treatment [9].

2.2 Surgical treatment

Surgical interventions are often endorsed for patients with complications of LUTS such as AUR, renal insufficiency, bladder calculi or .recurrent urinary tract infections, persistent gross hematuria secondary to BPH [14]. Further, other candidates for surgery includes the patients refractory to other medical management, or men with unacceptable side-effects following drug therapies and requested for active treatment [15].

Open prostatectomy, transurethral resection of the prostate (TURP), and transurethral incision of the prostate (TUIP) are some of the conventional surgical treatment options for symptomatic BPH. The removal of obstructing tissue was first achieved by open prostatectomy in early 1900s [16] and considered as gold standard for the surgical treatment, but later replaced by TURP. Significant improvement in LUTS were observed with TURP, and it takes only 20–30 min, to resect an average gland weighing 30 g. Though TURP is considered to be as the hallmark by the urologist, the one against which other surgical options are compared, but it carry the complications of excessive bleeding and longer hospital stay [16, 17]. TUIP a comparative less invasive technique than TURP and with similar improvements in symptoms is recommended for prostate gland weighing <25 g of the prostate [18]. An electrosurgical modification of the TURP and TUIP technique i.e. transurethral vaporization (TUVP), is reserved particularly for the patients with a small prostate and bleeding disorders. Its long term efficiency has been found to be comparable with that of TURP, but number of patients reported for irritative symptoms as side effects [19].

Raising the temperature of the cells through the use of low level radiofrequency (microwave) in prostate to 40-45°C (hyperthermia), 46-60°C (thermotherapy) and 61-75°C (transrectal thermal ablation) are found to be more specific techniques for the necrosis of obstructive tissue without affecting normal cells [16]. In comparison to high-energy TUMT with increased morbidity, low range TUMT has been found to well tolerated in patients with reasonable improvement in flow rate and less effect on sexual function [20]. Another simple, safe and relatively inexpensive technique to deliver high frequency radiowaves (temperature range 90-100°c) to produce localized necrotic lesions in hyperplastic tissue is Transurethral needle ablation (TUNA). Its a method of choice over TURP in younger men and with small sized gland, wishing to preserve sexual function, as it poses a low or no risk for incontinence and impotence [21, 22].

Laser vaporization or prostectomy, has been found to be another safe, effective and widely used form of MIT technique with significant improvement in urinary flow rates and symptoms. Light at different wavelength is being generated using four types of lasers, namely potassium titanyl phosphate (KTP) diode laser; neodymium: yttrium-aluminum-garnet (Nd: YAG) laser, and holmium YAG laser (Ho:YAG), that cause irreversible cellular damage, followed by their coagulation necrosis and ultimately vaporization of tissues. Further, evolution in holium laser prostatectomy i.e. Holmium laser enucleation of the prostate (HoLEP) is being used for prostate of all sizes at considerable faster rate than TURP and now considered to be as new gold standard for the treatment of BPH. HoLEP relieves the pressure on the urethra tube by anatomically enucleating the majority of excess benign prostate tissue. Short operative time & hospital stay, minimal blood loss and fluid absorption, and bladder neck contractures are some of the advantages of laser prostatectomy over the TURP and other conventional techniques [23–25].

2.3 Pharmacological treatment

The clinical manifestations of BPH are primarily precipitated by increased resistance to the flow of urine through the bladder neck and/or compressed prostatic urethra. Thereby, the treatment strategies are targeted to decrease the urinary resistance by reducing the prostatic volume. A number of strategies are available but great strides in the development of alpha-adrenergic blockers and anti-androgen (androgen deprivation therapy) have fueled this evolution.

2.3.1 Alpha adrenergic blockers

Alpha adrenergic blockers relaxes the smooth muscle in and around the prostate and bladder neck without affecting the detrusor muscle of the bladder wall thus relieve the obstruction due to dynamic component of LUTS. The rationale for this approach is based on that noradrenaline (NA) acts at alpha-1 adrenergic receptors (α_1-AR) in the neck and sphincter of the urinary bladder to promote contraction and urinary retention. NA also acts at alpha-1 adrenergic receptors to control the smooth muscles in the prostate capsule and urethra [26]. Prazosin with a piperazinyl quinazoline nucleus, was the first clinically investigated selective α_1-adrenergic receptor antagonist for BPH with 1000-fold greater affinity than that for α_2-receptor. But, because of associated important adverse effects like postural hypotension and retrograde ejaculation, soon it was withdrawn from market [27]. The next advancement in drug therapy was the advent of selective α_1-drugs, Terazosin and Doxazosin, structurally close analogs of Prazosin [28].

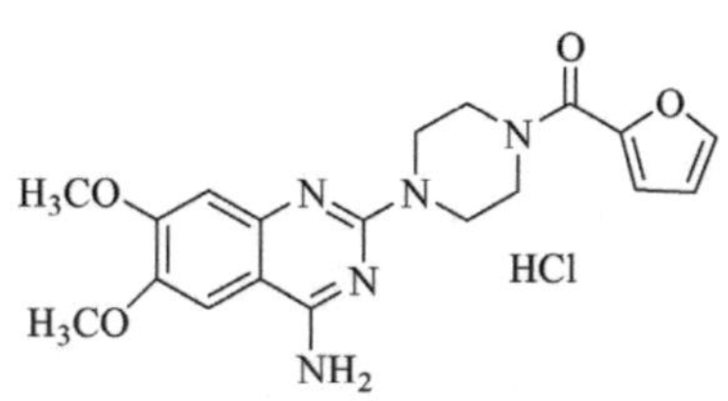

Prazosin

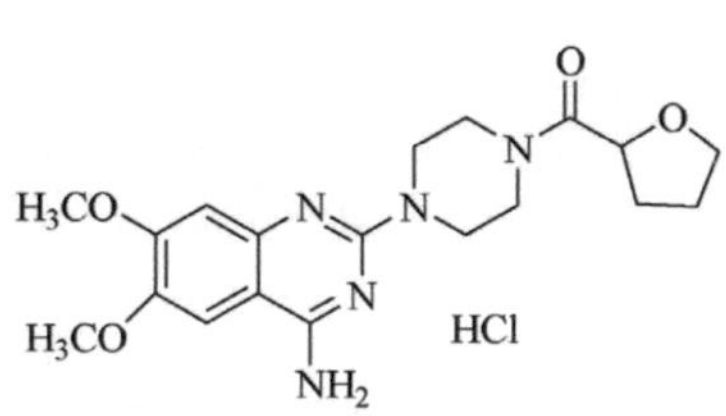

Terazosin

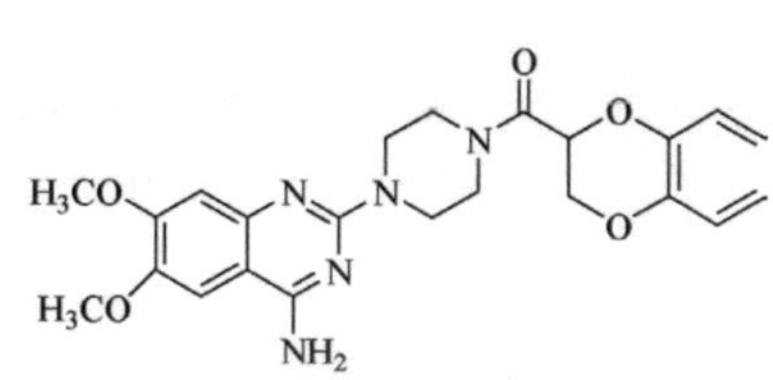

Doxazosin

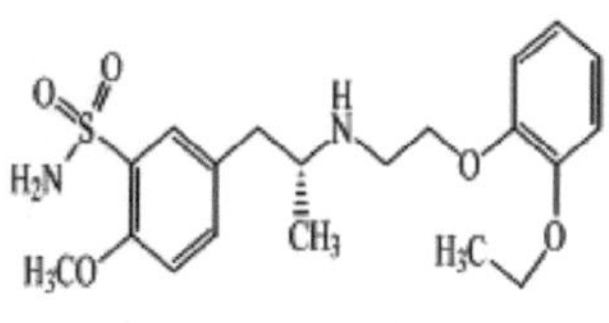

Tamsulosin

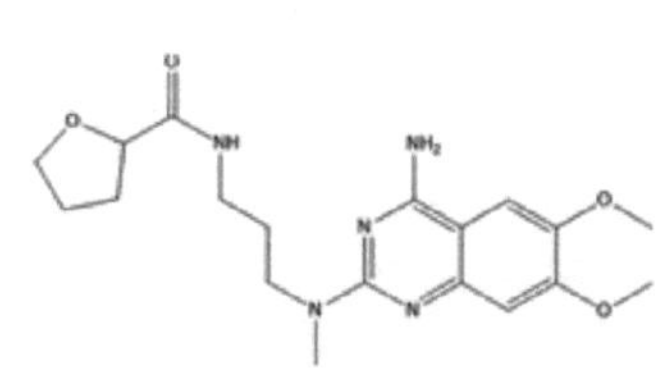

Alfuzosin

Molecular studies have further identified three subtypes α_{1A}, α_{1B} and α_{1D} of the α_1-AR. The α_{1A} is predominant in prostate, whereas α_{1B} subtype has been found to be predominant in blood vessels [29]. Their relative distribution and concentration in the bladder, prostate, neck, brain and vascular smooth muscle have been exploited to develop uroselective α_1-adrenergic antagonists with reduced side-effects. Tamsulosin was launched as the first subtype selective α_1-AR antagonist, but third uroselective α_1-AR antagonist with ten fold more selectivity for α_{1A}-receptor subtype compared to α_{1B}-receptor subtype. Whereas, Alfuzosin, with comparable clinical efficacy to that of tamsulosin was the fourth uroselective α_1-AR antagonist with almost similar affinity for all of the α_1 receptor subtypes and [12, 30]. Currently, Tamsulosin and Alfuzosin are the most widely prescribed medications as selective α_1-AR antagonists for the LUTS associated with BPH.

2.3.2 Androgen deprivation therapy

The biological basis of this therapy lies in the observation that the androgens (dihydrotestosterone). plays a crucial role in the development and maturation of prostate gland. Furthermore, BPH does not develop in the patient who are castrated prior to the puberty [31, 32]. Androgen suppression causes reduction in prostatic volume which is believed to decrease the considerable responsible static component of bladder outlet obstruction resulting from benign prostatic hyperplasia [33].

Progestational agents like medogesterone, and hydroxyprogesterone acetate, acts in reversible manner and are capable of decreasing testosterone level in the serum by inhibiting the release of luteinising hormone (LH) [34]. Further, desensitization and down regulation of pituitary gonadotropin releasing hormone (GnRH) receptors by agonistic GnRH analogues is well established approach in the clinical treatment of BPH [35]. These agents (leuprolide, and Nafarelin acetate) [36], results in the blockage of gonadotropin release from the anterior pituitary gland followed by the suppression of steroidal sex hormones production. Antiandrogens like flutamide, cyproterone acetate, curcumin analogues bicalutamide, 16 substituted/non-substituted D-homo-pregnane derivatives) compete for androgen receptor with the natural ligand (DHT) binding and are used therapeutically in BPH patients [37–41].

Plethora of the evidences has indicated the role of estrogen along with male androgens in the aging men with BPH condition. Estradiol is the product of the peripheral conversion of testicular and adrenal androgen in man under the influence of enzyme aromatase. Under the estrogenic effect, stromal and epithelial interactions presumably mediate and regulate the proliferative activity of the prostate [42]. Testolactone, atermestone, TZA-2237, and abiraterone are some of the aromatase inhibitors and found application in non-surgical treatment of BPH by blocking this peripheral conversion [43–45].

The importance of androgendeprivation by the use of antiandrogen agents was underscored by the fact that these centrally acting drugs decrease the testosterone level, and cause complications like erectile dysfunction and loss of libido [45, 46]. Therefore, search for the new drugs with more efficacies, selectivity and relative broader therapeutic index was being pursued and continued accrual resulted in the development of 5α-reductase inhibitors.

3. 5α-reductase inhibitors

5α-reductase (5AR) is a nuclear membrane bound enzyme that converts testicular endogenous testosterone T to dihydrotestosterone DHT in the presence of

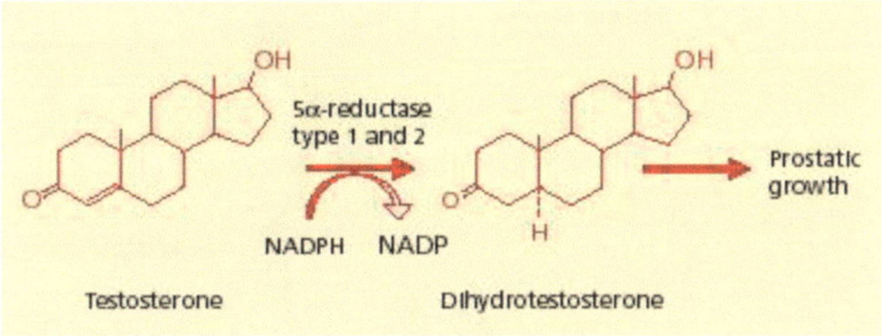

Figure 4.
5α-reductase enzymatic action.

cofactor NADPH. Thus 5AR dictates the cellular availability of DHT to prostatic epithelial cells and consequently modulate its growth as shown in **Figure 4** [47].

Thus, inhibiton of androgen action by 5α-reductase represents a logical treatment of 5α-reductase activity disorder i.e. BPH. 5ARI decreases the dihydrotestosterone concentration by blocking the enzyme and, provide relief from the symptoms related to the static mechanical obstruction caused by BPH by shrinking the size of prostate [48]. Further, the rationale for use of 5ARI is rooted in the observation that these agents are more specific to DHT action without affecting/lowering T level, thus capable of decreasing long term side effect of castration associated with loss of testosterone, without compromising the efficacy of hormonal therapy [49, 50].

3.1 Physiology of androgens release

Figure 5 indicating the control of testicular androgen production by hypothalamus and the pituitary gland. Neurons in preoptic area of the hypothalamus secrete the decapeptide lutenizing hormone releasing hormone (LHTH), in a pulsatile fashion, which in turn stimulates the release of lutenizing hormone (LH) from the pituitary. After reaches to the testies, LH binds to the high affinity receptor present on the surface of leydig cells and stimulate them to produce testosterone. Released T travels in the blood either in the free state or after binding with protein [43]. Circulating testosterone levels in a negative feed back mechanism regulates the secretion of hypothalamus and pituitary.

Major androgen in the adult male is Testosterone (T) and 98% of all T in the prostate is of testicular origin, whereas only 5–10% being produced by adrenal gland [51]. The unbound T diffuses into the prostate cell (target organ), where most of it gets converted to dihydrotestosterone (DHT) by the membrane bound NADPH dependent enzyme 5α-reductase. Within prostate, DHT binds to cytosol androgen receptor protein (AR) followed by entry of DHT-AR complex into the nucleus, where it stimulates the RNA synthesis after interacting with DNA binding sites (**Figure 6**) [52].

T and DHT differ in their physiological action and T also binds to androgen receptor but with lesser affinity to that of DHT [53]. According to Burckovsky and Wilson postulation T acts as a prohormone and DHT is found to be the main active hormone in androgen sensitive tissue [47]. With in embryo, T is responsible for the transformation of Wollffian ducts in epididymis, seminal vesicle & differential ducts & and responsible for production of DHT after activating the expression of 5AR. On the other hand, DHT in embryo is found to be crucial for the sexual differentiation of male foetus organ, formation of external genitilia, like urethera and prostate. After puberty, it's the T that determines the modification of external

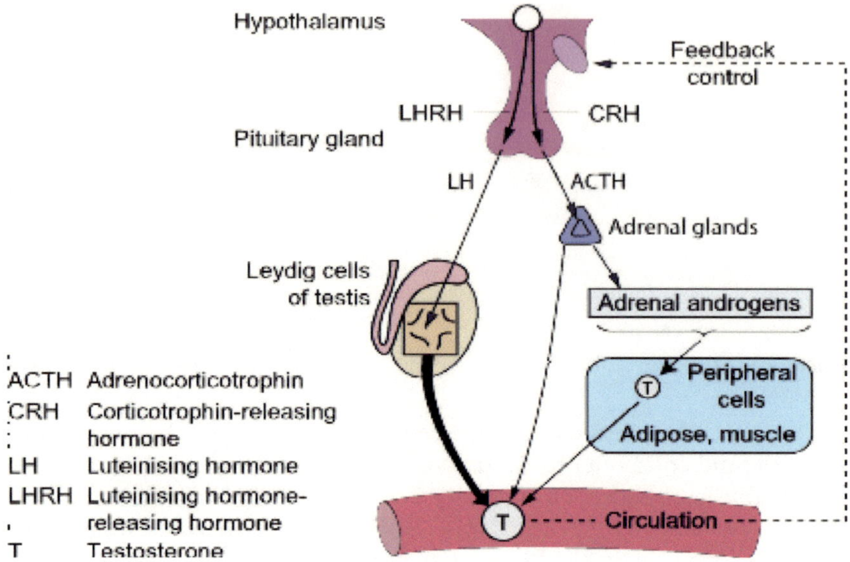

Figure 5.
Physiology of androgen release.

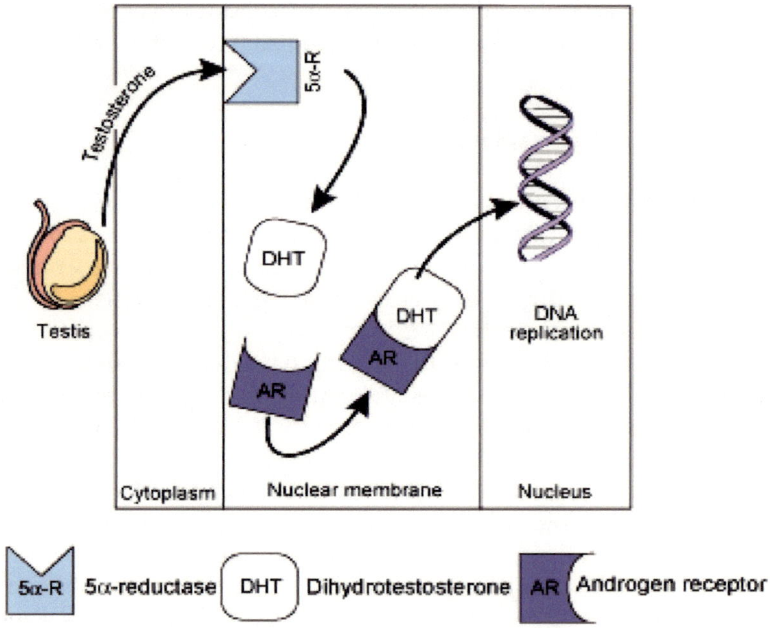

Figure 6.
Interaction of androgen within prostate cell.

genitilia, deeping of voice, increase of muscle mass, spermatogenesis, and male sexual behavior. In contrary to that DHT formation in male puberty is related with the increase of facial & body hair,r and the enlargement of prostate [32, 37, 54, 55].

Further crucial role of DHT was discovered through male pseudohermaphroditism, a genetic disorder with complete or partial 5α-reductase deficiency. Decreased 5AR activity not only resulted in low level of DHT [56, 57], but also accompanied by several distinguished features at the critical juncture of foetal and postnatal development [58]. Male with such condition showed ambiguous external genitilia at-birth [59], often raised as girls, little facial hairs as adults, no temporal receding hairline, small prostate no acne and normal libido. Whereas, female with 5AR deficiency did not show any clinical symptoms.

Excessive production of DHT is associated with development of several endocrine diseases such as acne, alopecia in men, male pattern baldness, hirusitism in women, prostatic carcinoma and benign prostatic hyperplasia [7]. In BPH, concentration of DHT is found to 2.5 fold higher than in normal prostate.

3.2 Isozyme of 5α-reductase

The family of 5AR is composed of three known isoenzymes with the types I and II being the most known. Steroidal 5α-reductase is a system of NADPH dependent enzymes that catalyzes the irreversible conversion of 4-en-3-oxo-steroid to the corresponding 5α-H-3-oxo-steroid [60–62]. Based on the anatomical location, biochemical properties, and tissue expression pattern three different isozymes of 5AR have been isolated, expressed and characterized (**Table 1**). The type 2 isozyme is predominantly present in the prostate, seminal vesicle, epididymis, genital skin, and liver. It has been found to be essential for differentiation of male external genitilia during foetal life, and its deficiency leads to the condition known as male pseudohermaphroditisms [63, 64]. Whereas, type 1 is not the major species expressed in the prostate and exhibit only micromolar affinities for steroidal natural substrate (T) [65, 66].

Both the isoforms have optimal activity at different pH range as type 1 is active at alkaline pH of 8.5, while type 2 is active at pH 4.7–5.5. Studies have shown that the activity of type I enzyme is several times higher in PC than in BPH. Whereas the 5AR type II (5AR-2) isoenzyme with higher affinity for T at the optimum pH 5.5 predominates in the prostate and other genital tissues and plays a major role in BPH [67, 68].

	5AR-1	5AR-2	5AR-3
Gene	SRD5A1	SRD5A2	SRD5A3
Location	5p15	2p23	4q12
Length (b)	36,173	56,385	25,458
Protein size	259	254	319
Transmembrane helices	5	4	6
Protein weight (Da)	29,459	28,393	36,521
Optimal pH	6–8.5	5–5.5	6.9
Affinity for testosterone	Km = 1.7 μM	Km = 0.2 μM	—
In vitro inhibition	$K_i \geq 300$ nM	$K_i = 3–5$ nM	—
Localization (in tissues)	Sebaceous glands of skin, sweat glands, dermal papilla cells, fibroblasts from all areas	Prostate, genital skin, epididymis, seminal vesicles	Hormone refractory prostate cancer cells, pancreas, brain, skin, adipose tissue
Selectivity to the inhibitors	Inhibitors with 4-methyl-4-aza functionality are very potent	4-aza, 6-aza and charged 3-substitutents derivatives are highly selective.	—

Table 1.
5AR isozymes and their characteristic features.

A new isoenzyme of 5AR, type III (5AR-3) have been identified recentlyin castration resistant prostate cancer (CRPC) cells as well as in other tissues such as pancreas, brain, skin and adipose tissues [69, 70]. The length, location and other characteristics of these isoenzymes have been presented in **Table 1** [71].

3.3 Mechanism of 5α-reductase action

The detailed chemical and kinetic mechanisms of conversion of T into DHT by 5AR have been investigated as follows:

3.3.1 Chemical mechanism

Figure 7 is indicating the proposed mechanism of T reduction to DHT under the influence of 5α-reductase. It is based on the known regio and stereoselectivity of the reduction that involves the formation of binary complex between the enzyme and NADPH, followed by formation of ternary complex with the substrate [72, 73]. Binary complex formation follows the activation of the enone system by based on its strong interaction with commonly present electrophilic residue (E^+) (proton, +ve charged group, proton donor) in the active site. Enone activation gives the delocalized carbocation which is being reduced selectively at C-5, on the α-face, by a direct hydride transfer from NADPH and lead to the formation of the enolate of DHT [74]. Generated intermediate duly coordinated with $NADP^+$ on the α- face, is further attacked by a proton on the β-face at C-4 and results into the formation of

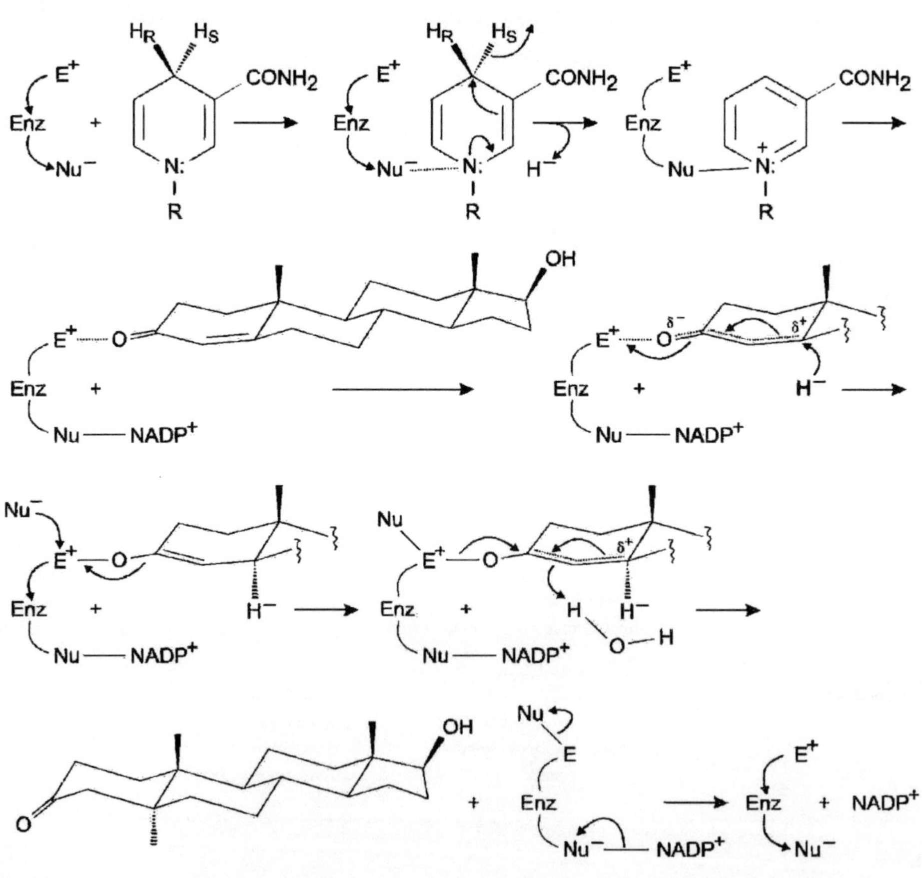

Figure 7.
Chemical mechanism of action of 5α-reductase.

ternary complex E-NADP$^+$–DHT. Towards the end of reaction, release of DHT gives the binary NADP$^+$-enzyme complex, followed by the release of NADP$^+$ leaving the enzyme free for further catalytic reactions.

3.3.2 Kinetic mechanism

The kinetic mechanism was studied for the natural substrate T using rat and human prostatic 5α-reductase and both the models showed similar kinetic mechanism as shown in **Figure 8.** 1,4-reduction of the substrate (T) depends on the initial velocity data from progesterone and 5α-reductase, wherein NADP$^+$ is found to be competitive versus NADPH but non-competitive versus progesterone. Further catalysis occurs with the initial release of DHT followed by NADP$^+$.

3.4 Classification of 5α-reductase inhibitors

The control of the physiological action of major androgen DHT, without significant change in the overall profile of other hormones especially (T), through the inhibition of specific enzyme 5AR involved in its synthesis and metabolism, plays an important role in the design of ARIs, mimicking the electronic and steric properties of the enolate [75].

The identification of different isozymes of 5AR, their specific role in physiological and pathological developments of BPH has opened the door for more specific and selective inhibitors of this enzyme [76]. Broadly 5α-reductase inhibitors have been divided into following major groups a) Transition state analogues b) Mechanism based inhibitors c) Structure based.

3.4.1 Transition state analogues

Based on chemical mechanism of 5AR, two possible transition states (**Figure 9**) have been postulated substrate like and product like. [77, 78]. The 'substrate like' transition state is the one in which the C-5 has not yet changed it sp^2-hybridization and the structure of C-3, C-4, and C-5 are similar to those of intermediate carbonation. On other hand in 'product like' TS C-5 has assumed its final sp^3 hybridization and structure of C-3, C-4 and C-5 are similar to those of enol form of DHT.

Transition state analogue states that the binding to the enzyme and thus its inhibition could be greater for molecules being mimic of the transition of the enzymatic process [77].

3.4.2 Mechanism based analogues

According to the kinetic mechanism of T reduction to DHT, three different types (Type A, B and C) of inhibitors have been identified [79, 80]:

a. Type A: Inhibitors compete with substrate testosterone and cofactor NADPH i.e. bisubstrate.

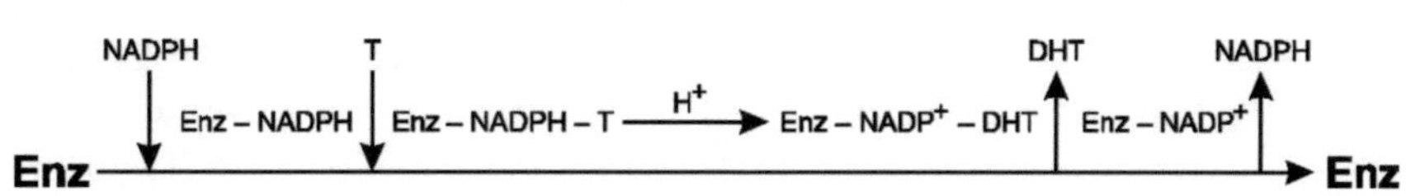

Figure 8.
Kinetic mechanism.

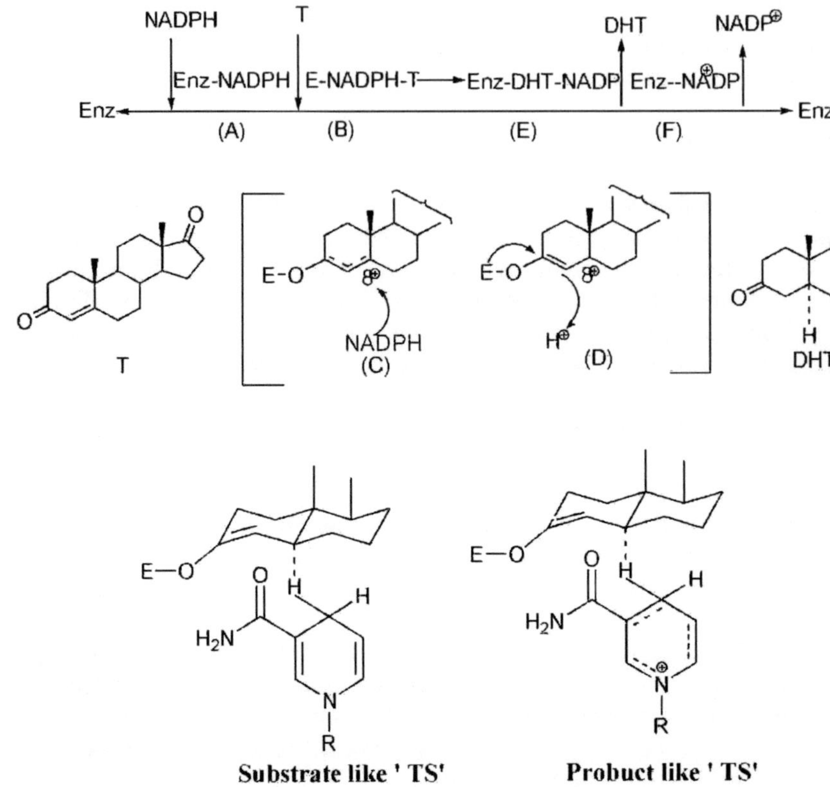

Figure 9.
Transition states of the enzyme (5AR).

b. Type B: These are the compounds that got the potential to bind reversibly to NADPH-enzyme complex and competitive with natural substrate T thus competitive inhibitors.

c. Type C: Such inhibitors fit the enzyme- NADP complex and are uncompetitive versus the substrates.

Number of steroidal and non-steroidal analogs ranging from classical, reversible and irreversible inhibitors, and transition state analogues to mechanism-based analogues have been synthesized and evaluated during last two decades as shown in **Figure 10**.

Biological basis for the steroidal inhibitors lies in the observation that enzyme could be best inhibited by the compounds having structural similarities to natural substrate i.e. T. One of the earlier report in 1970 by Voigt and Hsia, indicate the ability of 23 steroidal hormones to inhibit 5AR in human skin thus the efficacy of steroidal derivatives in BPH [81]. Progesterone, a competitive substrate of T, restrained transformation by upto 93.3% and was converted to 5-pregnane-3, 20-dione. Great affinity of progesterone for 5AR was further indicated by its high value of inhibitory constant (K_i = 700 nm). Other potent inhibitors were deoxycortisone, deoxycortisone acetate and dehydroepiandrosterone [82]. In 1973, synthesis and evaluation of series of 5ARI, indicated the key structural requirements for the 5ARI activity i.e. presence of 4-en-3-one function and 17β-side chain having one or more oxygen functionalities. Molecules possessing these features act as competitive inhibitors of 5AR, therefore, all of them could be regarded as a substrate of the enzyme 4-en-3-one steroids [83]. The clinically approved first inhibitor was

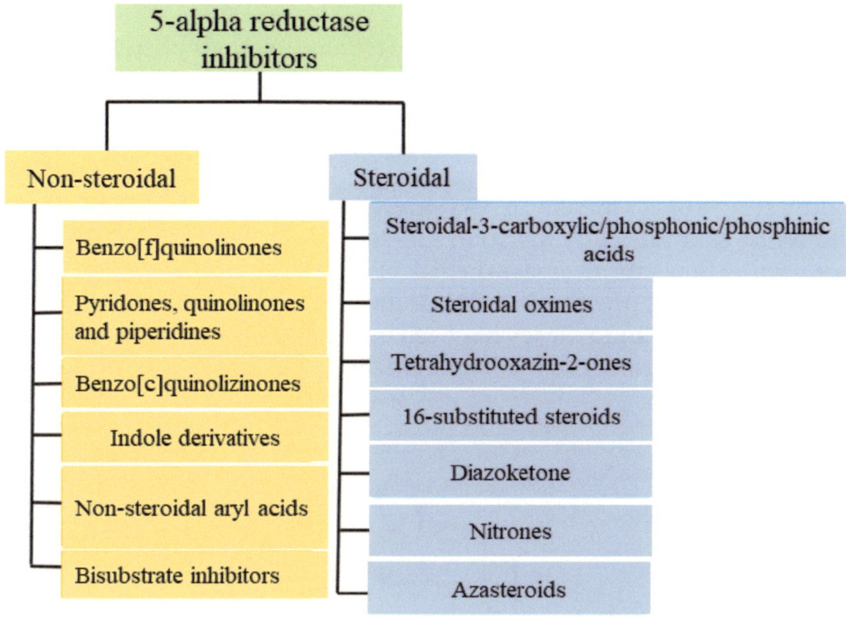

Figure 10.
Chemical classes of 5AR inhibitors.

prepared by modification of the structure of naturally existing substrates. This modification included the substitution of various hetero atom such as nitrogen, by forming the azasteroids by replacing carbon atom of the ring with nitrogen in the steroidal moieties.

Finasteride.

Chemically Finasteride (MK-906) is 17β-(N-tert-butyl-carbamoyl)-4-aza-5α-androst-1-en-3-one. It was synthesized in 1984, and got clinical approval in 1992 in the United States as the first 5α-reductase inhibitor for the treatment of BPH [84]. It is a competitive inhibitor of 5α-reductase type 2 with 10-fold high affinity than type 1 and forms a stable complex with enzyme. Clinical doses of 5 mg/day has been found to decrease the prostatic DHT level by 70 to 90%, in human beings, thus decreases prostate volume or size followed by improvement in urinary flow rate [85, 86]. It has neither any other hormone (androgenic, antiandrogenic) related properties, nor it interferes with the binding of T or DHT to the androgen receptor [87]. Though significant improvement in term of increased flow rates and decreased prostate-specific antigen level has been observed in finasteride-treated group. But, its long term usage results in common side effects like decreased libido, ejaculatory dysfunction, or impotence, while rashes and breast enlargement have also been observed in some of the patients.

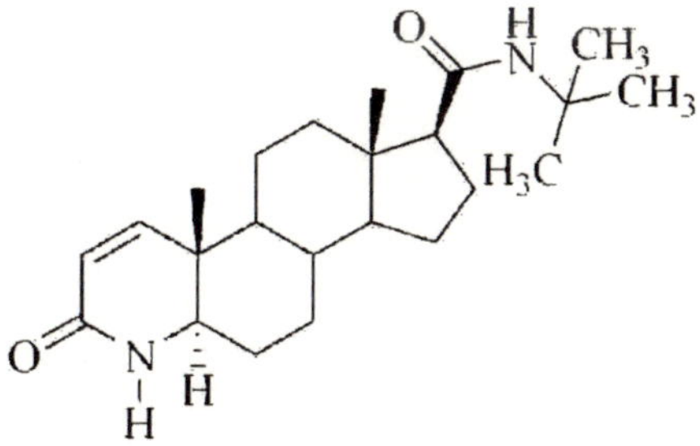

Finasteride

Dutasteride.

Chemically dutasteride is 17β-N-{2, 5-bis (trifluoromethyl) phenyl)} -3- oxo- 4-aza- 5α- androst - 1- ene - 17-carboxamide and belongs 4-aza-steroids [86] It was approved in 2002 by the US FDA for the symptomatic treatment of BPH. Unlike finasteride, dutasteride is a nonselective competitive inhibitor of both isozymes. 5α-reductase type 1 and type 2.

At clinical dose of 0.5 mg/day, it decreases DHT levels >90%, by forming a stable complex with a slow rate of dissociation constant. Dutasteride has been found to improve urinary flow rate, decrease the risk of AUR and need for surgery by reducing the size of enlarged prostate [88–90]. Dutasteride is found to be 60 times more active than finasteride and efficacy has been improved in terms of symptom score, maximal urinary flow rate, and quality of life [86].

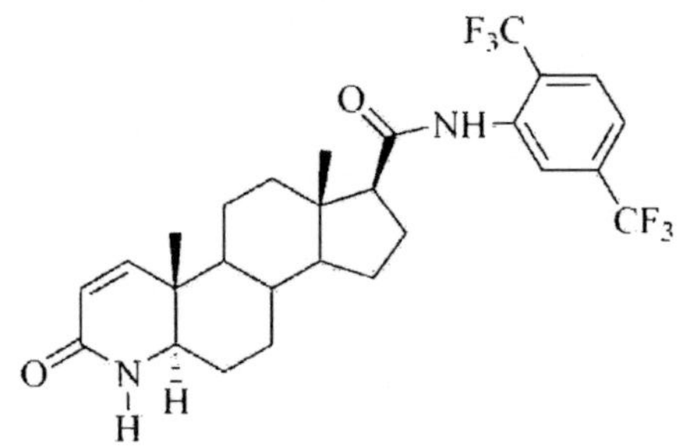

Dutasteride

These two drugs have been found to display competitive blocking effect in short-term kinetic, whereas long-term reaction analysis revealed their irreversible inhibitory effect by forming a stable complex of enzyme-bound intermediates [91]. The binding affinity between 5α-R isoenzyme and 4-azasteroids can be described in the two step mechanism:

$$E + I \underset{}{\overset{K_i}{\rightleftharpoons}} EI \xrightarrow{K_3} EI^*$$

Where K_i is the inhibition constant for the first step equilibrium and K_3 is the rate constant for the time-dependent second step [92]. Mechanistically, finasteride has been proven to be 5α-R2 inhibitor by acting on alternative substrate for 5α-R2 which is initially bound to highly stable complex of enzyme-bound NADP-dihydrofinasteride. The resulting adduct is finally processed to form dihydrofinasteride [93]. The bisubstrate complex of NADP-dihydrofinasteride is a potent inhibitor with dissociation constant $k_i 1x\ 10^{-31}$ M that makes it as one of the extremely potent known non-covalently bound complexes.[180,183] Finasteride is also

	K_3 (s^{-1})	**K_1 (IC_{50}, nM)**	**K_3/K_1 (M^{-1} s^{-1})**
5α-R 1			
Finasteride inhibition	1.4×10^{-3}	360	4×10^{3}
Dutasteride inhibition	1.1×10^{-3}	6	1.8×10^{5}
5α-R 2			
Finasteride inhibition	2.2×10^{-2}	69	3.2×10^{5}
Dutasteride inhibition	4.9×10^{-3}	7	6.8×10^{5}

Table 2.
Inhibition of 5α-R isozymes by clinically approved drugs.

known for its inhibitory effect on 5α-R1. However, the resultant dihydrofinasteride complex has comparatively lower rate constant (**Table 2**).

4. Combination therapy

The scientific rationale for combining 5ARIs and α_1-AR antagonists is based on their different and complementary modes of action, that help in managing static and dynamic component responsible for of an enlarged prostate gland and symptoms of LUTS. The rationale for this combination was further recommended on account of rapid relief of symptoms by the α_1-AR antagonists, without targeting the underlying disease process along with mid or more sustained relief of symptoms by the 5ARIs [94]. The efficacy and safety of the treatment with different combinations versus treatment with either agent alone has been investigated by different groups in large mulitcentral trials [95, 96].

Veterans Affairs Cooperative Study and Prospective European Doxazosin group evaluated the combination of finasteride with terazosin & doxazosin, respectively for one year. Significant improvement in the symptom score and flow rate was observed with α_1-AR antagonists alone or combination therapy as compared to placebo or finasteride alone, but there was no significant difference observed for combination therapy over α_1-AR antagonists alone. Short term successful trials were followed by studying the combination of finasteride and doxazosin for a period of 4.5 years as Medical therapy for Prostate Symptoms. Finasteride alone and this particular combination reduced the risk of AUR and need for BPH-related surgery versus placebo, whereas none of these outcomes were reduced significantly in patients consuming doxazosin alone.

Outcomes of another long term study examining the role of combination of dutasteride and tamsulosin (CombAT) over the α_1-AR antagonists (tamsulosin) alone would be a major step in assessing the combination therapy and treatment decision [97]. Though present observations demonstrated a higher incidence of impotence with combination therapy compared with 5ARIS, in addition to higher incidence of α_1-AR antagonists-mediated dizziness, hypotension [93]. Cost-effectiveness studies by Nickel suggest that the combination therapy is more suitable for men at high risk for BPH progression, patients with high symptom score, large prostate volume and low q_{max} value.

Author details

Neelima Dhingra
Pharmaceutical Chemistry, MNASc, MPASc, MIABMS, University Institute of Pharmaceutical Sciences, Panjab University, Chandigarh, India

*Address all correspondence to: neelimad08@gmail.com

References

[1] Kenny, B.; Ballard, S.; Blagg, J.; Fox, D.Pharmacological Options in the Treatment of Benign Prostatic Hyperplasia. *J. Med. Chem.* **1997**, *40*, 1293-1315

[2] Arora RP, Nayak RL, Malhotra V, Mohanty NK, Kulkarni SK. Role of herbal drugs in the management of benign prostatic hyperplasia: Clinical trial to evaluate the efficacy and safety of Himplasia. *Med Update* **2003**;11:55-58.

[3] Thorpe A, Neal D. Benign prostatic hyperplasia. Lancet **2003**; 361: 1359-1367.

[4] Djavan B, Remzi M, Erne B, Marberger M. The pathophysiology of benign prostatic hyperplasia. *Drugs Today* **2002**; 38: 867-870.

[5] Culig Z, Hobisch A, Cronauer MV, Radmayr C, Hittmair A, Zhang J, et al. Regulation of prostatic growth and function by peptide growth factors. *Prostate* **1996**; 28: 392-405.

[6] Geller, J. Benign prostatic hyperplasia: pathogenesis and medical therapy*J. Am. Geriatr. Soc.* **1991**, *39*, 1208-1216.

[7] Jenkins EP, Andersson S, Imperato-McGinley J, Wilson JD, Russell DW. Genetic and pharmacological evidence for more than one human steroid 5-alpha-reductase. J Clin Invest **1992**; 89: 293-300.

[8] Giacomo, N.; Antonio, G.; Mario, G.; Vincenzo, F. Critical review of guidelines for BPH diagnosis and treatment strategy. *Europ. Urol. Suppl.*, **2006**, 5(4), 418-429.

[9] Rosette, J.; De, L.; Alivizatos, G.; Madersbacher, S.; Rioja Sanz, C.; Nordling, J.; Emberton M. Guidelines on benign Prostatic hyperplasia. *Europ. Assoc. Urol.*, **2002**, 5-54.

[10] Ziada, A.; Rosenblum, M.; Crawford, E.D. Benign prostatic hyperplasia: An overview. *Urology,* 1999, 53(3), 1-6.

[11] Madsen, F.A.; Rhodes, P.R.; Bruskewitz, R.C. Reproducibility of pressure-flow variables in patients with symptomatic benign prostatic hyperplasia. Urology, **1995**, 46(6), 816-820

[12] Griffiths, D.J. Pressure-flow studies of micturition. *Urol. Clin. North. Am.*, **1996**, 23(2), 279-297.

[13] Eri, L.M.; Tveter, K.J. Treatment of benign prostatic hyperplasia. a pharmacoeconomic perspective. Drugs Aging, **1997**, 10(2), 107118

[14] Orandi, A. transurethral resection versus transurethral incision of the prostate. *Urol. Clin. Am.*, **1990**, 17(3), 601-612.

[15] Curtis Nickel, J.; Carlos, E.; Thomas, F.; Whelan Ryan, F. Paterson, Hassan Razvi. 2010 Update: Guidelines for the management of benign prostatic hyperplasia. *Can. Urol. Assoc.* J., **2010**, 4, 310-316.

[16] Garg, G.; Singh, D.; Saraf, S. Management of benign prostate hyperplasia: An overview of α-adrenergic antagonist. *Biol. Pharm. Bull.*, **2006**, 29, 1554-1558.

[17] Blohm, T.R.; Laughlin, M.E.; Benson, H.D.; Johnston, J.O.; Wright, C.L.; Schatzman, G.L. Pharmacological induction of 5-alpha reductase deiciency in rats: Separation of testosterone-mediated and 5-alpha-dihydrotestosterone-mediated effects. *Endocrinology*, **1986**, 119, 959-966.

[18] Reid, P.; Kantoff, P.; Oh, W. Antiandrogens in prostate cancer. Invest. New Drug, **1999**, 17, 271-284

[19] Wolf, H.; Madson, P.O. Treatment of benign prostatic hypertrophywith progestational agents: A preliminary report. *J. Urol.*, **1968**, 99, 980-985.

[20] Mearini, L.; Massimo, P. Transrectal high-intensity focused ultrasound for the treatment of prostate cancer: Past, present and future. *Indian J. Urol.*, **2010**, 26, 4-11.

[21] Griesinger, G.; Felberbaum, R.; Diedrich, K. GnRH-antagonists in reproductive medicine. *Arch. Gynecol. Ob*stet., **2005**, 273, 71-78.

[22] Scott, W.W.; Wade, J.C. Medical treatment of benign nodular prostatic hyperplasia with cyproterone acetate. J. Urol., **1969**, 101, 81-85.

[23] Ai, N.; De Lisle, R.K.; Yu, S.J.; Welsh, W.J. Computational model for predicting the binding afinity of ligands for the wild type androgen receptor and a mutated variant associated with human prostate cancer. *Chem. Res. Toxicol.*, **2003**, 16, 1652-1660

[24] Ito, K.; Fukabori, Y.; Shibata, Y.; Suzuki K.; Mieda, M.; Gotanda, K. Effects of new steroidal aromatase inhibitor, TZA-2237 and/or chlormadinone acetate on hormone-induced and spontaneous canine benign prostatic hyperplasia. *Eur. J. Endocrinol.*, **2000**, 143, 543-54.

[25] Handratta, V.D.; Vasaitis, T.S.; Njar, V.C.; Gediya, L.K.; Kataria, R.; Chopra, P. Novel C-17-heteroaryl steroidal CYP17 inhibitors/ antiandrogens: Synthesis, in vitro biological activity, pharmacokinetics and antitumor activity in the LAPC4 human prostate cancer xenograft model. *J. Med. Chem.*, **2005**, 48, 2972-2984

[26] Dhingra, N.; Bhagwat, D. Benign prostatic hyperplasia: An overview of existing treatment. *Ind. J. Pharmacol.*, **2011**, 43(1), 612

[27] Donovan, J.L.; Peters, T.J.; Neal, D.E.; Brookes, S.T.; Gujral, S.; Chacko, K.N. A randomized trial comparing transurethral resection of the prostate, laser therapy and conservative treatment of men with symptoms associated with benign prostatic enlargement. J. Urol., **2000**, 164, 65-70.

[28] Kumar, V.L; Wahane, V.D. Current status of 5α-reductase inhibitors in the treatment of benign hyperplasia of prostate. Indian J. Med. Sci., **2008**, 62, 167-175.

[29] Ramon, J.; Lynch, T.H.; Eardley, I.; Ekman, P.; Frick, J.; Jungwirth, A. Transurethral needle ablation of the prostate for the treatment of benign prostatic hyperplasia: A collaborative multicentre study. Br. J. Urol., **1997**, 80, 128-134

[30] Dyrstad, S.W.; Shah, P.; Rao, K. Chemotherapy for prostate cancer. Curr. Pharm. Res, **2006**, 12, 819-837.

[31] Siiteri, P. K.; Wilson, J. D. Dihydrotestosterone in prostatic hypertrophy. I. The formation and content of dihydrotestosterone in the hypertrophic prostate of man. *J. Clin. Invest.* **1970,** *49*, 1737-1745.

[32] Blohm, T. R.; Laughlin, M. E.; Benson, H. D.; Johnston, J. O.; Wright, C. L.;Schatzman, G. L.; Weintraub, P. M. Pharmacological Induction of 5 Alpha-Reductase Deficiency in the Rat: Separation of Testosterone-Mediated and 5 Alpha-Dihydrotestosterone-Mediated Effects. *Endocrinol.* **1986,** *119* (3), 959-966.

[33] Reid, P.; Kantoff, P.; Oh, W. Antiandrogens in Prostate Cancer. *Invest. New Drugs.* **1999,** *17* (3), 271-284

[34] Wolf, H.; Madsen, P. O. Treatment of Benign Prostatic Hypertrophy with Progestational Agents: A Preliminary Report. *J. Urol.* **1968,** *99* (6), 780-785

[35] Griesinger G, Felberbaum R, Diedrich K. GnRH-antag onists in reproductive medicine. *Arch Gynecol Obstet,* 2005,273(2):71-78

[36] Peters, C. A.; Walsh, P. C. The Effect of Nafarelin Acetate, a Luteinizing-Hormone Releasing Hormone Agonist, on Benign Prostatic Hyperplasia. *N.Engl. J. Med*. **1987**, *317* (10), 599-604.

[37] Scott, W. W.; Wade, J. C. Medical Treatment of Benign Nodular Prostatic Hyperplasia with Cyproterone Acetate. *J. Urol*. **1969,** *101* (1), 81-85

[38] Caine M, Perlberg S, Gordon R (1975) The treatment of benign prostatic hypertrophy with flutamide: a placebo controlled study. *J Urol* 114:564-568

[39] Ai, N.; DeLisle, R. K.; Yu, S. J.; Welsh, W. J. Computational Models for Predicting the Binding Affinities of Ligands for the Wild-Type Androgen Receptor and a Mutated Variant Associated with Human Prostate Cancer.*Chem. Res. Toxicol*. **2003**, *16* (12), 1652-1660.

[40] Ohtsu, H.; Xiao, Z.; Ishida, J.; Nagai, M.; Wang, H. K.; Itokawa, H.; Su, C. Y.; Shih, C.; Chiang, T.; Chang, E.; Lee, Y.; Tsai, M. Y.; Chang, C.; Lee, K. H. *J. Med. Chem.* **2002**, *45*, 5037.

[41] Bratoeff, E. A.; Herrera, H.; Ramirez, E.; Solorzano, K.; Murillo, E.; Quiroz, A.; Cabeza, M. Antiandrogenic effect of 16-substituted, non-substituted and D-homopregnane derivatives. *Chem. Pharm. Bull. (Tokyo)* **2000**, *48*, 1249-1255.

[42] Ito, K.; Fukabori, Y.; Shibata, Y.; Suzuki, K.; Mieda, M.; Gotanda, K.; Honma, S.; Yamanaka, H. Effects of a new steroidal aromatase inhibitor, TZA-2237, and/or chlormadinone acetate on hormone-induced and spontaneous canine benign prostatic hyperplasia. *Eur. J. Endocrinol.* **2000**, *143*, 543.

[43] Schweikert, H. U.; Tunn, U. W. Effects of the aromatase inhibitor testolactone on human benign prostatic hyperplasia.*Steroids* **1987**, *50*, 191-200.

[44] Henderson, D.; Habenicht, U. F.; Nishino, Y.; Kerb, U.; el Etreby, M. F. *J. Steroid Biochem.* **1986**, *25*, 867.

[45] Guess, H. A.; Heyse, J. F.; Gormley, G. J. The Effect of Finasteride on Prostate-Specific Antigen in Men with Benign Prostatic Hyperplasia. *Prostate.* **1993**, 22 (1), 31-37

[46] John, I. J.; Oncol, J. H. 5-Alpha reductase inhibitors and the treatment of benign prostatic hyperplasia. *Drugs of Today* **1993**, *29*, 335-346.

[47] Bruchovsky, N.; Wilson, JD. The Intranuclear Binding of Testosterone and 5-alpha-androstan-17-beta-ol-3-one by Rat Prostate. *J Biol Chem.* **1968**, *25,* 5953-5960

[48] Andriole, G.; Bruchovsky, N.; Chung, L. W.; Matsumoto, A. M.; Rittmaster, R.; Roehrborn, C.; Russell, D.; Tindall, D. Dihydrotestosterone and the prostate: the scientific rationale for 5alpha-reductase inhibitors in the treatment of benign prostatic hyperplasia. *J. Urol.* **2004**, *172*, 1399-1403.

[49] Kurup, A.; Garg, R.; Hansch, C. Comparative QSAR analysis of 5α-reductase inhibitors.*Chem. Rev.* **2000**, *100*, 909-911

[50] Faragalla, J.; Bremner, J.; Brown, D.; Griffith, R.; Heaton, A. Comparative pharmacophore development for inhibitors of human and rat 5-alphareductase. *J. Mol. Graph. Model.* **2003**, *22*, 83-92

[51] Trachtenberg, J.; Hicks, L. L.; Walsh, P. C. Androgen- and estrogen-receptor content in spontaneous and experimentally induced canine prostatic

hyperplasia *J. Clin. Invest.* **1980**, *65*, 1051-1059

[52] Liao, S. Cellular receptors and mechanisms of action of steroid hormones *Int. Rev. Cytol.* **1975**, *41*, 87-172

[53] Wilson, E. M.; French, F. S. Binding properties of androgen receptors.*J.Biol.Chem.* **1976**, *251*, 5620-5629.

[54] Quemener, E.; Amet, Y.; di Stefano, S.; Fournier, G.; Floch, H. H.; Abalain, J. H. Purification of testosterone 5 alpha-reductase from human prostate by a four-step chromatographic procedure*Steroids* **1994**, *59*, 712-718.

[55] Kelce, W. R.; Lubis, A. M.; Braun, W. F.; Youngquist, R. S.; Ganjam, V. K. Influence of rete testis fluid deprivation on the kinetic parameters of goat epididymal 5 alpha-reductase. *Steroids* **1990**, *55*, 27-31

[56] Imperato-McGinley, J.; Peterson, R. E.; Gautier, T.; Sturla, E. **Male pseudohermaphroditism secondary to 5a-reductase deficiency - a model for the role of androgens in both the development of the male phenotype and the evolution of a male gender identity.** *J. Steroid Biochem.* **1979**, *11*, 637-645.

[57] Griffin, J. E.; Wilson, J. D. The syndromes of androgen resistance.*N. Engl. J Med.* **1980**, *302*, 198-209.

[58] Imperato-McGinley, J.; Guervo, L.; Gautier, T. Imperato-McGinley J, Guerrero L, Gautier T, Peterson RE: Steroid 5α-reductase deficiency in man: An inherited form of male pseudohermaphroditism. *Sciences* **1994**, *186*, 1213-1215.

[59] Peterson, R. E.; Imperato-McGinley, J.; Gautier, T.; Sturla, E. Peterson RE, Imperato-McGinley J, Gautier T, Sturla E. Male pseudohermaphroditism due to steroid 5-alpha-reductase deficiency. *Am. J. Med.* **1977**, *62*, 170-191.

[60] Anderson, K. M.; Liao, S. Selective retention of dihydrotestosterone by prostatic nuclei. *Nature* **1968,** *219*, 277-279.

[61] Voigt, W.; Fernandez, E. P.; Hsia, S. L. Transformation of testosterone into 17 beta-hydroxy-5 alpha-androstan-3-one by microsomal preparations of human skin *J. Biol.Chem.* **1970**, *245*, 5594-5599.

[62] Monslave, A.; Blaguies, J. A. Partial characterization of epididymal 5α reductase in the rat. *Steroids* **1977**, *30*, 41-51.

[63] Andersson, S.; Berman, D. M.; Jenkins, E. P.; Russell, D. W. Deletion of steroid 5 alpha-reductase 2 gene in male pseudohermaphroditism *Nature* **1991**, *354*, 159-161

[64] Labrie, F.; Sugimoto, Y.; Luu-The, V.; Simard, J.; Lachance, Y.; Bachvarov, D.; Leblanc, G.; Durocher, F.; Paquet, N. Structure of human type II 5 alpha-reductase gene. *Endocrinology* **1992**, *131*, 1571-1574.

[65] Harris, G.; Azzolina, B.; Baginsky, W.; Cimis, G.; Rasmusson, G. H.; Tolman, R. L.; Raetz, C. R.; Ellsworth, K. Identification and selective inhibition of an isozyme of steroid 5a-reductase in human scalp. *Proc. Natl. Acad. Sci. U S A* **1992**, *89*, 10787-10791

[66] Normington, K.; Russell, D. W. Tissue distribution and kinetic characteristics of rat steroid 5 alpha-reductase isozymes. Evidence for distinct physiological functions. *J. Biol. Chem.* **1992**, *267*, 19548-19554

[67] Li, J.; Ding, Z.; Wang, Z.; Lu, J. F.; Maity, S. N.; Navone, N. M., Kim, J. Androgen Regulation of 5α-Reductase Isoenzymes in Prostate Cancer: Implications for Prostate Cancer

Prevention. PLoS One, **2011**, 6 (12), e28840.

[68] Uemura, M.; Tamura, K.; Chung, S.; Honma, S.; Okuyama, A.; Nakamura, Y.Novel 5α-Steroid Reductase (SRD5A3, Type-3) is Overexpressed in Hormone-Refractory Prostate Cancer. *Cancer Sci.* **2008**, 99, 81-86.

[69] Paba, S.; Frau, R.; C, Godar, S.; Devoto, P.; Marrosu, F.; Bortolato, M. Steroid 5α-reductase as a novel therapeutic target for schizophrenia and other neuropsychiatric disorders. *Curr. Pharm. Des*. **2011**, 17 (2), 151-167

[70] Lao, K.; Sun, J.; Wang, C.; Lyu, W.; Zhou, B.; Zhao, R.; Xiang, H. Design,Synthesis and Biological Evaluation of Novel Androst-3, 5-diene-3-carboxylic Acid Derivatives as Inhibitors of 5α-Reductase Type 1 and 2. *Steroids.* **2017**,124, 29-34

[71] Henderson, D.; Habenicht, U.F.; Nishino, Y.; Kerb, U.; El Etreby, M. F.Aromatase Inhibitors and Benign Prostatic Hyperplasia. J. Steroid Biochem.**1986**, 25 (5), 867-876.

[72] Holt, D. A.; Levy, M. A.; Matcalf, B. W. *Advances in Medicinal Chemistry*; Maryanoff, B. E., Maryanoff, C. A. Eds.; JA Press Inc: Greenwich, Connecticut, **1993**, p 1.

[73] 179 Abell, A. D.; Henderson, B. R. Steroidal and non-steroidal inhibitors of steroid 5α-reductase. *Curr.Med.Chem* **1995**, *2*, 583-597

[74] Brian, W.; Metcalf, K. J.; Joseph, P. B. Synthesis of 3-keto-4-diazo-5-α-dihydrosteroids as potential irreversible inhibitors of steroid 5-α-reductase. *Tetra.Lett.* **1980**, *21*, 15-18

[75] Foley, C. L.; Bott, S. R.; Shergill, I. S.; Kirby, R. S. An update on the use of 5alpha-reductase inhibitors. *Drugs of Today* 2004, *40(3)*, 213-223.

[76] Frye, S. V. Inhibitors of 5-alpha reductase . *Curr.Pharm.Design* **1996**, *2*, 59-84

[77] Bull, H. G.; Garcia –Calvo, M.; Andersson, S.; Baginsky, W. F.; Chan, H. K.; Ellsworth, D. E.; Milles, R. R.; Stearns, R. A. Mechanism-Based Inhibition of Human Steroid 5α-Reductase by Finasteride: Enzyme-Catalyzed Formation of NADP–Dihydrofinasteride, a Potent Bisubstrate Analog Inhibitor. *J.Amer.Chem.Soc.* **1996**, *118*, 2359-2365

[78] Guarna, A.; Marriace, A.; Danza, G.; Seirio, M. *"Sex Hormones and Antihormones in Endocrine Dependent Pathology Basic and Clinical Aspects"*; Motta, M., Seirio, M. Eds.; Elsevier Sciences: Amsterdam, **1994**.

[79] Liang, T.; Heiss, C. E.; Ostrove, S.; Rasmusson, G. H.; Cheung, A. Binding of a 4-Methyl-4-Aza-Steroid to 5α-Reductase of Rat Liver and Prostate Microsomes. *Endocrinology* **1983**, *112*, 1460-1468

[80] Erb, J. M.; Heaslip, J. I.; Brandt, M. Potent inhibition of human steroid 5alpha-reductase (EC 1.3.1.30) by 3-androstene-3-carboxylic acids. *Bioorg. Med.Chem.* **1989**, *17*, 372-376

[81] Alauddin, M.; Martin-Smith, M. Biological Activity in Steroids Possessing Nitrogen Atoms. *J. Pharm. Pharmacol*. **1962**, 14 (1), 469-495.

[82] Singh, H.; Jindal, D. P.; Yadav, M. R., Kumar, M. Heterosteroids and Drug Research. *Prog. Med. Chem*. **1991**, 28, 233-300

[83] Martínez, M. D.; Edelsztein, V. C.; Durán, F. J.; Di Chenna, P. H., Burton, G. Synthesis of 6-Azaprogesterone and 19-Hydroxy-6-Azasteroids. *Steroids*. **2013**, 78 (1), 34-37.

[84] Tian G, Mook RA, Moss ML, Frye SV. Mechanism of

time-dependent inhibition of 5 alpha-reductase by delta 1-4-azasteroids: Toward perfection of rates of time-dependent inhibition by using ligand-binding energies. *Biochemistry* **1995**;34:13453-3459.

[85] Harris GS, Kozarich JW. Steroid 5α-reductase inhibitors in androgen-dependent disorders. *Curr Opin Chem Biol* **1997**;1:254-259.

[86] Kumar VL, Wahane VD. Current status of 5α-reductase inhibitors in the treatment of benign hyperplasia of prostate. *Indian J Med Sci* **2008**;62:167-175.

[87] Rittmaster RS, Lemay A, Zwicker H, Capizzi TP, Winch S, Moore E, *et al.* Effect of finasteride, a 5 alpha-reductase inhibitor, on serum gonadotropins in normal men. *J Clin Endocrinol Metab* **1992**;75:484-488.

[88] Marihart S, Harik M, Djavan B. Dutasteride: a review of current data on a novel dual inhibitor of 5-alpha reductase. *Rev Urol* **2005**;7:203-210.

[89] Miller J, Tarter TH. *Update on the use of dutasteride in the management of benign prostatic hypertrophy.* Clin Interv Aging **2007**;2:99-104.

[90] Roehrborn CG. The clinical benefits of dutasteride treatment for LUTS and BPH. *Rev Urol* **2004** ;6: S22-30.

[91] Schmidt, L. J.; Tindall, D. J. Steroid 5 α-reductase inhibitors targeting BPH and prostate cancer. *J. Steroid Biochem. Mol. Biol*. **2011,** 125, 32-38.

[92] Ludwig, P.; Holzhütter, H. G.; Colosimo, A.; Silvestrini, M. C.; Schewe, T.; Rapoport, S. M. A kinetic model for lipoxygenases based on experimental data with the lipoxygenase of reticulocytes. *Eur. J. Biochem*. **1987**, 168, 325-337.

[93] Salvador, J. A.; Pinto, R. M.; Silvestre, S. M. Steroidal 5α-reductase and 17αhydroxylase/17, 20-lyase (CYP17) inhibitors useful in the treatment of prostatic diseases. *J. Steroid Biochem. Mol. Biol*. **2013**, 137, 199-222.

[94] Tanguay S, Awde M, Brock G, Casey R, Kozak J, Lee JJ, et al. Diagnosis and management of benign prostatic hyperplasia in primary care. *Can Urol Assoc J* **2009**;3:S92-S100.

[95] Curtis NJ. BPH: Costs and treatment outcomes. *Am J Manag Care* **2006**;12:S141-8

[96] *Bullock TL, Andriole GL. Emerging drug therapies for benign prostatic hyperplasia. Expert* Opin Emerg Drugs **2006**;11:111-23

[97] Miller, J.; Tarter, T.H. Combination therapy with dutasteride and tamsulosin for the treatment of symptomatic enlarged prostate. *Clin. Interv. Aging*, **2009**, 4, 251-258.

Chapter 6

Monoamine Oxidase A (MAO-A): A Therapeutic Target in Lung Cancer

Chandreyee Datta, Sukhamoy Dhabal and Ashish Bhattacharjee

Abstract

Monoamine oxidase-A (MAO-A), a pro-oxidative enzyme catalyzes the oxidative deamination of endogenous and exogenous monoamines/neurotransmitters like dopamine, serotonin, norepinephrine or tyramine and converting them into their corresponding aldehydes and reactive oxygen species (ROS). Hyperactivity of MAO-A has been shown to be involved in depression, neuro-degeneration including Parkinson's and Alzheimer's diseases, neuropsychiatric disorders and cardiovascular diseases. Our recent results however demonstrated the involvement of MAO-A in promoting aggressiveness of lung carcinoma. We found both constitutive and inducible expression of MAO-A in non-small cell lung cancer cells H1299 and in A549 lung epithelial carcinoma cells. By using knockout (by CRISPR-Cas9 gene editing technology) or knockdown (using MAO-A specific esiRNA) MAO-A cells we demonstrated the role of MAO-A in promoting lung cancer aggressiveness and epithelial to mesenchymal transition (EMT). From our observations, we can conclude that MAO-A may be considered as a potential therapeutic target for the intervention and treatment of lung carcinoma.

Keywords: monoamine oxidase-A (MAO-A), non-small cell lung carcinoma, 15-lipoxygenase, metastasis, epithelial to mesenchymal transition (EMT)

1. Introduction

Monoamine oxidase A (MAO-A) is a mitochondrial outer membrane-bound enzyme that catalyzes oxidative deamination of biogenic amines and subsequently generates reactive oxygen species (ROS) in the form of hydrogen peroxide (H_2O_2) as a catalytic by product. It is widely present in almost all mammalian cell types except in erythrocytes [1, 2].

It is well documented that abnormalities of MAO-A levels and activity can lead to neuropsychiatric disorders as it plays a vital role in the regulation of neurotransmitters. Moreover, MAO-A hyperactivity has been shown to be associated with depression and previous reports implicate MAO-A inhibitors as effective therapeutics against clinical depression and anxiety [3, 4]. Involvement of MAO-A has also been shown in neurodegenerative diseases including Parkinson's and Alzheimer's disease by inducing oxidative stress-mediated apoptosis [5, 6]. MAO-A deficiency and abnormal activity has also been associated with impulsive aggressive behavior [7], neuropsychiatric disorders [1], pancreatic beta cell function [8] and glucose

 IntechOpen

metabolism [9]. In addition to neurodegenerative disorders and neuroinflammatory diseases, mounting evidences have been suggested about the contribution of MAO-A in cardiovascular diseases like myocardial injury [10], heart failure [11], cardiac cell apoptosis [12] etc.

Previously it has been shown that MAO-A has a major contribution in the resolution of inflammation and thus been reported as a signature marker of alternatively activated monocytes/macrophages [13]. Reactive oxygen species (ROS) can predispose cancer cells to DNA damage and cause tumor initiation and progression [14]. This suggests that MAO-A might have a significant role in cancer. Rybaczyk et al. [15] carried out a study where they have analyzed a subset of cancer datasets concentrating on genes involved in the serotonergic pathway. Genechip datasets consisting of cancerous tissue from human, mouse, rat, or zebrafish were obtained from the GEO database [16, 17]. Initially, obvious changes that were common in various types of cancers were identified by comparing gene expression between cancerous tissues and normal tissues for each type of cancer. This study strongly demonstrated that MAO-A suppression could be linked to increased risk of cancer and MAO-A expression was decreased in 95.4% of human cancer patients and 94.2% of animal cancer cases compared to the non-cancerous controls [15].

In contrast, high Gleason grade or poorly differentiated prostate cancer exhibited increased MAO-A expression [18], and the increased level of MAO-A promoted prostate cancer metastasis [19, 20]. Furthermore, it was also reported that the overexpressed MAO-A in prostate cancer cells was the main causative agent for the reduced expression of E-cadherin and increased expression of vimentin and Twist at both mRNA and protein levels in prostate cancer [19]. These studies suggested a likely role of MAO-A for the progression of prostate cancer by mediating EMT. However, contradicting results were reported in case of HCC [21] and cholangiocarcinoma [22]. Therefore, it can be hypothesize from these reports that MAO-A functions across different cancer cells in a context specific manner and so, it is essential to further uncover the function of MAO-A in other cancers. All these reports indicate towards the emerging role of MAO-A in tumor growth, migration and metastasis. However, none of these studies revealed the role of MAO-A in lung cancer growth, migration and metastasis and its mechanistic regulation.

MAO-A can be present constitutively in many different types of cancer cells (like in H1299 lung cancer, HCT116 colorectal cancer or in LNCap prostate cancer cells), or it can be induced by Th2 cytokines IL-13/IL-4 in A549 lung epithelial carcinoma cell line or monocytic U937 cell line. In a very recent study by our group we showed that IL-13 mediated induction of MAO-A in human bronchial epithelial cell NHBE as well as in human lung carcinoma cell line A549. We also explored the mechanisms involved in the regulation of the expression/activity and function of MAO-A during IL-13-induction and presented evidence that Stat6, 15-LO and PPARγ are the critical regulators that are involved in regulating MAO-A gene expression and activity of A549 cells which further demonstrated the concerted mechanistic effects of these genes during IL-13-activation. Altogether, the IL-13/STAT6, STAT3, STAT1/15-LO/PPARγ signaling axis for regulating MAO-A gene expression and function add novel insights into the resolution of inflammation and in the progression of lung cancer [23]. In addition to that, our recent unpublished observation revealed that MAO-A plays an important role in regulating cancer cell aggressiveness and EMT transition. Moreover, a very recent report from Huang et al. has provided evidence that MAO-A plays a key role in EMT and HIF-1α protein accumulation induced by HPV-16 E7 in NSCLC cells, suggesting that MAO-A may be a potential therapeutic target for HPV-related NSCLC [24]. So, based on these recent findings, we have focused on the potential contribution of MAO-A in lung cancer aggressiveness and metastasis and EMT transition. Previous literature and our observations thus indicated that MAO-A could serve as a potential therapeutic target of lung cancer intervention and treatment.

In this chapter, we precisely highlighted the contribution of MAO-A in lung cancer aggressiveness and EMT and thus finally to the prognosis of the lung cancer patients primarily based on the observations obtained from our recent publication and ongoing research work and background studies. We also tried to explore the fact that MAO-A being a well-known contributor in neurological and neuropsychiatric disorders, how it could also be a tempting target of lung cancer treatment.

2. MAO-A is co-induced with 15-LO in human lung cancer cells during IL-13 activation

15-lipoxygenase (15-LO) is a lipid peroxidating enzyme which is substantially induced in human peripheral blood monocytes after IL-4/IL-13 activation. This enzyme oxidizes polyunsaturated fatty acids like linoleic and arachidonic acids to their corresponding hydroperoxides like 13-S-HPODE and 15-S-HPETE [23, 24], which have been implicated as inflammatory mediators in cell development and in the pathogenesis of various diseases [25–29].

In a very recent study, our group demonstrated that MAO-A is co-induced with 15-LO in monocytes/macrophages, normal human bronchial epithelial (NHBE) cells and in A549 lung epithelial carcinoma cell line in response to IL-13 treatment [23]. Moreover, concordant 15-LO and MAO-A induction following IL-13 stimulation was also investigated in other lung epithelial cancer cells like in H1299 NSCLC. In H1299 cells, 15- LO expression level was very low and was not induced by IL-13. In contrast, MAO-A was constitutively present in H1299 cells but was not further induced upon IL-13 stimulation. These results thus demonstrate that in H1299 NSCLC, MAO-A is already overexpressed and no further IL-13-dependent induction occurs in these cells.

3. Transcriptional regulation of MAO-A and 15-LO upon IL-13 induction

In previous reports, it was demonstrated that Stats (Stat1, Stat3 and Stat6) are required for controlling IL-13- mediated 15-LO and MAO-A gene expression [30, 31]. Along with the same line, our recent research article further confirmed the direct binding of Stat transcription factors (Stat1, Stat3 and Stat6) to their cognate DNA binding sites present in the 15- LO promoter after IL-13 stimulation in primary monocytes [23]. On the other hand, predicted transcription factor binding sites of MAO-A promoter does not show any Stat consensus sequence in its promoter but reveals presence of a bunch of different other transcription factor binding sites like Sp1, GATA, TBP and GRE [32]. To justify and validate the prediction of different transcription factor binding sites located in the MAO-A promoter, we pursued experiment and presented evidence that in response to IL-13 stimulation, Sp1 transcription factor directly binds to the cognate DNA binding sites in the MAO-A promoter after IL-13 stimulation in primary monocytes [23]. After establishing the role of Sp1 trascription factor in regulation of MAO-A activity, now we are trying to explore the role of other transcription factors like GATA, TBP, GRE in the regulation of MAO-A.

It was previously demonstrated by our group that as Egr-1 and CREB binding sites are present in 15-LO promoter, in case of primary human monocytes, Egr-1 and CREB explicitly bind to their cognate sequences upon IL- 13 induction [33].

In addition to that, our data also affirmed that there are two distinct diverged imitate signaling pathways downstream of the IL-13 receptor that regulate 15-LO gene expression in primary monocytes [33]. Our study revealed that other than the conventional IL-4Rα-Jak2-Stat3-dependent pathway [34], there exists a IL-13Rα1-Tyk2-mediated pathway which is vital for IL-13-induced Egr-1 and CREB activation via MEK-ERK1/2. So, transcription factors Egr-1 and CREB plays a very crucial role in IL-13-induced 15-LO gene expression in primary human monocytes [33]. Based on these previous reports, and as 15-LO and MAO-A genes are co-induced upon exposure to IL-13 during alternative activation of monocytes, we further asked a question that whether Egr-1 and CREB transcription factors have any role in regulating IL-13- stimulated MAO-A gene expression in primary monocytes. Our experimental data strongly supported the regulatory role of transcription factors Egr-1 and CREB in mediating IL-13-stimulated MAO-A gene expression in primary monocytes probably by inducing 15-LO expression.

Therefore, collectively, the results presented previously by our group and in our recent study strongly suggest that Stats (Stat1, Stat3 and Stat6) as well as Egr-1 and CREB transcription factors are critical regulators of MAO-A activity in alternatively activated monocytes by IL-13.

Considering the fact that both 15-LO and MAO-A are co-induced upon IL-13 stimulation in primary human monocytes/macrophages and in A549 cells, and since activity of both of them are regulated by the transcriptional activation of several Stats (Stat1, Stat3 and Stat6), Egr-1 and CREB in monocyte/macrophage and in A549 cells, we further intended to explore whether one of them is the upstream regulator of another. As expected, experimental results the fact that IL-13- induced gene expression in lung carcinoma cell line A549 15-LO dependent [23].

4. IL-13-induced MAO-A contributes in lung cancer cell aggressiveness

It was previously reported that increased expression of MAO-A is associated with high grade aggressive prostate cancer [18, 35]. The ability of MAO-A to induce EMT in prostate cancer cells results in increased migratory, proliferative, invasive and metastatic potential through an elevation of ROS [19]. MAO-A-generated ROS modulates HIF1α (a master regulator of hypoxia) activity by suppressing PHD (oxygen dependent prolyl hydroxylases) activity. It was further verified that MAO-A enzymatic activity rather than the protein expression which is responsible for enhanced level of migration, invasion and proliferation of prostate cancer cells by the induction of EMT. These results thus suggest that MAO-A expression in high-grade tumors might have a likely role in maintaining a dedifferentiated phenotype and promoting aggressive behavior. Based on these previous reports, our group also investigated the contribution of MAO-A in enhancing the aggressiveness of different type of cancer cells like H1299 lung cancer cells and in HCT116 colorectal cancer cells where MAO-A is constitutively expressed and is responsible for promoting migration and invasion of these cancer cells (unpublished observations by our group). Based on our current observations, we also hypothesized that inducible MAO-A expression in A549 cells by IL-13/IL-4 might be involved in ling cancer progression and metastasis.

15-LO products HPODE/HPETE are well known PPARγ ligands and in our recent study, we have confirmed that IL-13 expression of MAO-A in lung epithelial carcinoma cell line A549 is dependent upon PPARγ [23]. Moreover, in our recent study, we have further confirmed that IL-13 stimulated A549 cell migration is mediated by MAO-A and requires both 15-LO and PPARγ activity. *In vitro* transwell migration assay using the MAO-A activity inhibitor Moclobemide along with the 15-LO inhibitor PD146176 and PPARγ antagonist GW9662 confirmed the above observation. In presence of Moclobemide, IL-13-induced A549 cell migration was reduced substantially whereas Moclobemide alone in absence of IL-13 showed no change compared to the unstimulated control, thereby suggesting a plausible role of MAO-A in A549 tumor cell migration *in vitro*. Similarly, significant reduction of migration was also observed in IL-13-activated A549 cells after treatment with

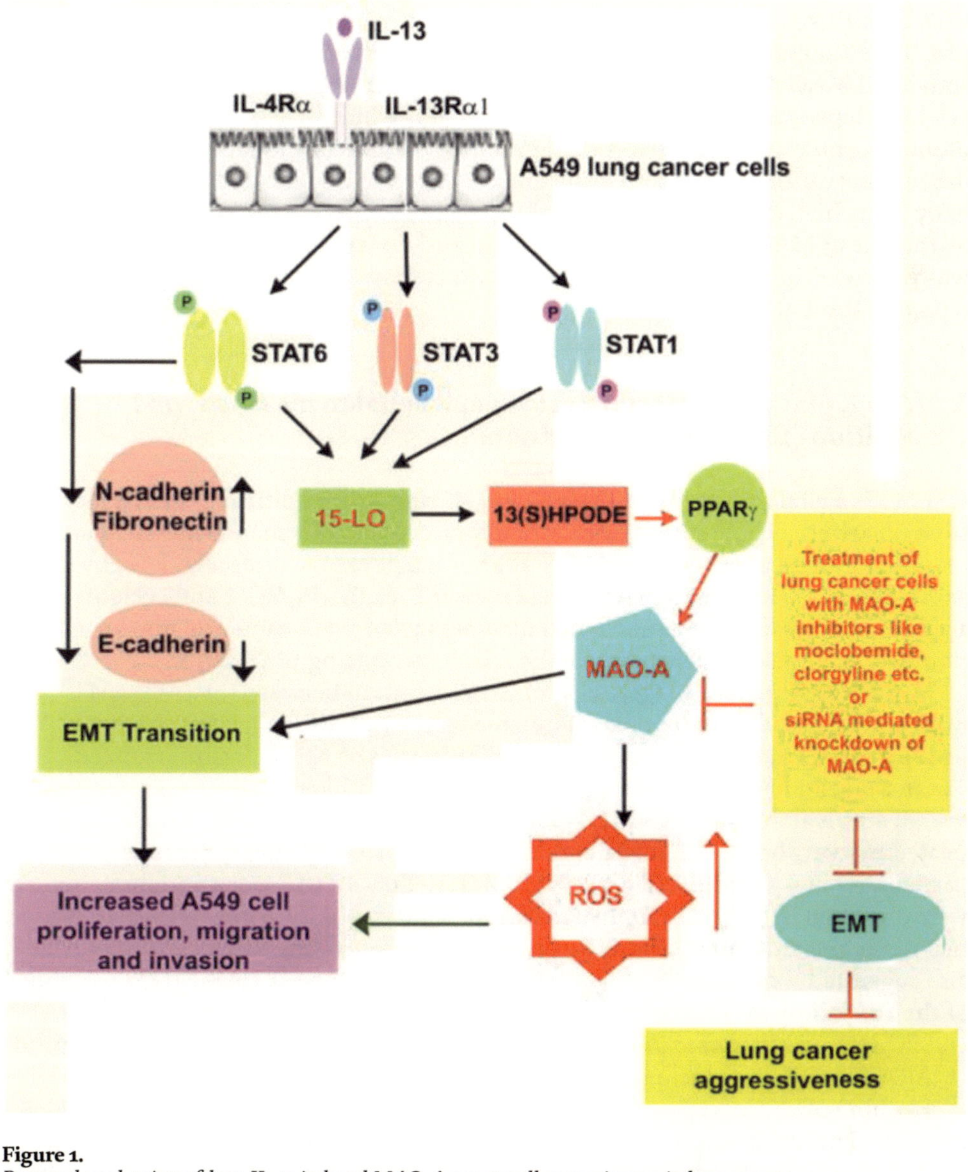

Figure 1.
Proposed mechanism of how IL-13 induced MAO-A cancer cell aggressiveness in lung cancer.
IL-13-mediated activation of MAO-A expression/activity in A549 lung epithelial carcinoma cells is activated by IL-13/ (STAT6, STAT3, STAT1)/15-LO/PPARγ signaling axis and expression/activity of this induced MAO-A mediates ROS generation which is believed to have significant role on A549 cell migration, invasion and proliferation which are all associated with the aggressiveness of this particular cancer cell. Moreover, treatment of A549 cells with MAO-A specific inhibitors like moclobemide or clorgyline or esiRNA- mediated knockdown of MAO-A gene in A549 cells showed significant downregulation in the migration, invasion and proliferation of lung cancer cells. This finding marked MAO-A as a promising therapeutic target for aggressive lung cancer treatment. IL-13: Interleukin 13, MAO-A: Monoamine oxidase A, STAT6: Signal transducer and activator of transcription 6, 15-LO: 15 lipoxygenase, PPARγ: Peroxisome proliferator-activated receptor gamma.

the 15-LO activity inhibitor PD146176 and PPARγ antagonist GW9662 in a dose-dependent manner compared to IL-13-stimulated positive control [23]. Moreover, we further confirmed the role of MAO-A in lung cancer cell migration and invasion in either MAO-A knockdown cells (using MAO-A esiRNA specific gene silencing) or MAO-A knockout cells (by using CRISPR-Cas9 gene editing technology) [Unpublished observations by our group].

Recently it was reported that the Th2 cytokine IL-13 contributes crucially in promoting EMT and enhancing aggressiveness (migration and invasion) in colorectal cancer (CRC) cells (HT29 and SW480 cells by triggering IL-13/IL-13Rα1/STAT6/ZEB1 signaling axis) [36]. Recent findings from our group further suggest that IL-13/IL-13Rα1/Stat6 signaling axis is involved in regulating the expression/activity of MAO-A in A549 lung epithelial carcinoma cells via a 15-lipoxygenase (15-LO)-dependent process involving PPARγ [23] which may be the main reason of promoting migratory, invasive and metastatic potential of the cancer cells (unpublished observations). But these results further need to be validated in *in-vivo* model or by using high grade lung cancer patient tissue samples to conclusively comment on the role of MAO-A on cancer progression and metastasis. The mechanism by which MAO-A lung cancer cell aggressiveness is described under a schematic representation in **Figure 1**.

5. MAO-A plays an important role in epithelial to mesenchymal transition (EMT) in lung carcinoma

EMT is a vital and commutative process, during which epithelial cells transit from polarized, cobble stone like cells to migratory, spindle-shaped mesenchymal cells. Apart from the morphological changes, changes at the molecular level by losing expression of various epithelial markers such as E-cadherin, ZO-1 and occludin, and acquiring expression of mesenchymal markers including N-cadherin, vimentin, and fibronectin are also very common in the cells experiencing EMT [37, 38].

Various studies have suggested a robust correlation between different EMT markers like E-cadherin, hypoxia inducible factor 1α (HIF-1α), twist, snail and poor prognosis in lung cancer [39]. Specially, in case of NSCLC, expression of Twist, Slug, and Foxc2 was identified as important marker of recurrence-free and overall survival in stage I NSCLC [40]. High expression pattern of various EMT related markers have been identified in advanced primary lung cancer specimens, particularly in squamous cell carcinoma [41]. Reduced EMT markers expression were observed in case of brain metastasis to primary NSCLC, supporting the notion that disseminated tumor cells undergo EMT at the site of metastasis [42, 43]. It was also suggested that enhanced expression of Forkhead box M1 (FOXM1), a member of the Fox family of factors, may have prognostic value for patients with NSCLC, and FOXM1 was shown to promote metastasis by inducing EMT through activation of the AKT/p70S6K signaling axis [44].

Plethora of earlier reports suggested that MAO-A-mediated generation of excessive intracellular level of hydrogen peroxide, a major ROS species can induce epithelial to mesenchymal transition (EMT) in cancer cells. An extensive study by Wu et al. in 2014 [19] elaborately described how MAO-A affects prostate cancer cell (PCa) growth and metastasis and demonstrated for the first time, that MAO-A induces EMT and augments hypoxic responses to increase the migratory, invasive, and metastatic potential of PCa cells.

In a recent study, Liu et al. [45] determined the expression of MAO-A and different EMT markers in 45 pairs of NSCLC and matched non-tumor adjacent lung tissues to further explore the connection between MAO-A expression and the EMT or the development of clinicopathological characteristics. From the results it was observed that both the protein and mRNA expression levels of MAO-A in NSCLC tissues were higher than those observed in the matched non-tumor adjacent lung tissues. Furthermore, in correlation with the previous notion, the enhanced expression of MAO-A in NSCLC tissues was positively associated with N-cadherin, Slug, and Twist, but negatively with E-cadherin expression. Furthermore, the elevated MAO-A expression in NSCLC tissues was also related with late stage NSCLC ($Z = -2.596$, $P = 0.029$) and lymph node metastases ($Z = -2.378$, $P = 0.020$). These findings indicated that MAO-A may have a role in inducing NSCLC progression by mediating EMT.

Next, considering the fact that high expression of monoamine oxidase A (MAO-A) in non-small cell lung cancer (NSCLC) is related to epithelial-mesenchymal transition (EMT) and the development of clinicopathological features of NSCLC, very recently, Yang et al. [46] tried to evaluate the role of a previously synthesized MAO-A inhibitor (G11) on inhibiting paclitaxel resistant NSCLC metastasis and growth. Experimental results showed that G11 significantly abrogated the viability of paclitaxel (PTX)-resistant NSCLC cell lines (A549/PTX and H460/PTX). G11 also abrogated the expression of MAO-A in A549/PTX and H460/PTX cells, which displayed relatively high MAO-A expression levels. Moreover, G11 was found to impede A549/PTX and H460/PTX cell migration and invasion. Furthermore, the *in-vivo* study also suggested that the co-administration of G11 and paclitaxel significantly suppressed tumor metastasis in H460/PTX lung metastasis models.

Considering these reports, we checked the expression of different EMT related markers like E-cadherin, N-cadherin, twist, snail, vimentin, etc. in MAO-A esiRNA treated A-549 cells as well as MAO-A knockout A549 cell line (by using CRISPR-Cas9 gene editing technology). As expected, these data confirmed regulatory role of MAO-A in EMT in lung carcinoma [Our unpublished observations]. Now to further validate the role of MAO-A in EMT and cancer metastasis, we are trying to explore the status of MAO-A, 15-LO and different EMT markers in lung cancer patient samples.

So, collectively, mounting evidences from different research reports and our recent observation strongly recommend further investigations for MAO-A as a tempting therapeutic target for lung cancer treatment.

6. Discussion

Monoamine oxidase A, an enzyme responsible for the oxidative deamination of biogenic amines, is well-known to be closely associated with impulsive aggressive behavior, anxiety, depression, and is considered as an indicator of psychological status [47, 48]. Recently, several studies have been focusing on the relationship between MAO-A expression and cancers [18–20, 49]. Initially, increased MAO-A expression was reported in high-grade aggressive prostate cancer, it was also demonstrated that increased expression of MAO-A was capable of mediating prostate tumorigenesis and metastasis [18–20]. Recently, MAO-A was reported as a novel decision maker in apoptosis and autophagy processes occurring within hormone refractory neuroendocrine prostate cancer cells [29]. Moreover, clorgyline, a known MAO-A inhibitor, was found to display anti-oncogenic and pro-differentiation effects on high-grade prostate cancer cells [50]. In contrary, MAO-A inhibitor-near-infrared dye conjugate was reported to reduce prostate tumor growth [51]. These findings suggest a likely role of MAO-A in mediating prostate cancer progression. However, in contrast to the previous reports of prostate cancer cells, Li et al. demonstrated that MAO-A expression was appreciably downregulated in clinical

HCC tissue samples [18], and MAO-A subdued HCC metastasis by hindering the adrenergic system and its transactivation of EGFR signaling [21]. Huang et al. also found that MAO-A expression was impeded by coordinated epigenetic and IL-6-driven events in human cholangiocarcinoma [22], and that overexpression of MAO-A suppressed cholangiocarcinoma growth and invasion [22].

So, from the above discussion it is clear that MAO-A level is regulated in different cancer cell lines in context specific manner. Moreover, MAO-A expression in high grade tumors may play a crucial role in promoting aggressive behavior of cancer cells. MAO-A degrades monoamine neurotransmitters by oxidative deamination and produces ROS. Increased level of ROS generation can be an important inducer of tumorigenesis, progression and metastasis in high grade cancers. Hence enhanced level of MAO-A expression and aggressive behavior of cancer cells may be correlated in advanced grade of cancer. From the studies on different type of cancer, it is quite evident that MAO-A may serve as a diagnostic biomarker and can also be applied as a therapeutic target in the treatment of cancer.

Furthermore, in case of NSCLC, as we have discussed earlier, research article by Liu et al. [45] supported our findings that MAO-A expression was significantly increased in NSCLC tissues, which was positively associated with EMT, late stages and lymph node metastases of the cancer, thus supporting the notion that MAO-A may play a role in NSCLC progression by regulating the EMT process.

In addition to that, along with the same line, Yang et al. [46] very recently established the role of MAO-A in lung cancer cell metastasis and EMT transition. So, these findings strongly recommend MAO-A as a promising therapeutic target of lung cancer treatment.

In our recent report, our result has demonstrated that IL-13- induced A549 cell migration was significantly downregulated in presence of moclobemide, a indicating a role of moclobemide in regulating lung cancer cell aggressiveness [23]. Wang et al. [52] that targeting MAO-A with FDA approved antidepressants could be a promising treatment option for the prostate cancer. It was reported by them that the antiandrogen enzalutamide (Enz) has improved survival in castration resistant prostate cancer (CRPC) patients. However, most patients eventually develop Enz resistance inducing by the androgen receptor (AR) splicing variant 7 (ARv7). Experimental results demonstrated that elevated expression of monoamine oxidase-A (MAO-A) is correlated with positive ARv7 detection in CRPC patients upon Enz treatment. Targeting MAO-A with phenelzine or clorgyline, the FDA-approved drugs for antidepression, resensitize the Enz resistant (EnzR) cells to Enz treatment and further subdudes EnzR cell growth *in-vitro* and *in-vivo*.

Moreover, Lee et al. [53] in 2013 have demonstrated that in case of LnCaP-LN3 prostate cancer cells MAO-A inhibitor pargyline significantly induced cell cycle arrest at the G1 phase compared to the control cells. In addition, pargyline induced an increase in the cell death rate by promoting apoptosis. Clinical depression is a very common feature in prostate cancer and mounting evidences have suggested that MAO-A levels are frequently elevated in different cancer types and MAO-A inhibitors (which are basically antidepressants) can serve as repurposing drugs for the treatment of cancer. Zarmouh et al. have identified a novel flavonoid MAO-A inhibitor which shows antiproliferative effect on prostate cancer cells [54].

Our recent research and different other research articles, it is well established that selective inhibitors of MAO-A like moclobemide, clorgyline, pargyline or a novel synthetic flavonoid can efficiently reduce cancer cell aggressiveness by either inhibiting cancer cell migration, proliferation or promoting apoptosis. So, selective inhibitors of MAO-A could also be used as a promising therapeutic agents for lung cancer treatment.

Moreover, in our unpublished observation, esiRNA specific knockdown of MAO-A in lung cancer cell A549 have shown significant downregulation of cancer cell migration, invasion and EMT transition. So, siRNA-mediated gene silencing approach for MAO-A could also be used as a potential therapeutic approach in lung cancer treatment.

7. Conclusion

In cancer cells, migration and invasion are the basic steps that control metastasis which is a principle cause of cancer-related death. Recent reports demonstrate that MAO-A is involved in promoting prostate cancer progression by inducing epithelial to mesenchymal transition (EMT) which ultimately causes set up of ROS, thus increasing the ability of migration and invasion of these cells. MAO-A enzymatic activity is shown to be the main causative agent for MAO-A-driven ROS generation in cancer cells which acts as the critical regulator of MAO-A-mediated functions like migration, invasion and proliferation.

Evidences from different reports and our own findings suggest that elevated levels of MAO-A is a key feature of advanced stage of lung carcinoma and plays a crucial role in lung cancer cell aggressiveness through induction of epithelial to mesenchimal transition. Thus it is really novel to report that an oxidative enzyme which is preliminarily found to be involved in depression and antisocial behavior now can be considered as a biomarker for lung cancer therapy.

Altogether, our results strongly support that MAO-A can be used as a potential therapeutic target of lung cancer treatment for the better prognosis of the disease.

Acknowledgements

The authors would like to acknowledge all the lab members for their support during this work and finally would like to thank the Department of Biotechnology, NIT Durgapur for extending the facilities for the research work. The unpublished research work was supported by Ramaligaswami Fellowship (to AB). by the Department of Biotechnology, Government of India, Research Grant SERB/SR/SO/HS-093/2013 from SERB (to A.B) Government of India and the research grant by the Department of Higher Education, Science and Technology, Biotechnology, Govt of West Bengal (to A.B). C.D was supported by DBT-RA fellowship and SD was supported by CSIR-SRF fellowship.

Author details

Chandreyee Datta, Sukhamoy Dhabal and Ashish Bhattacharjee*
Department of Biotechnology, National Institute of Technology, Mahatma Gandhi Avenue, Durgapur, 713209, India

*Address all correspondence to: ashish15lo@yahoo.com

References

[1] Shih JC, Chen K, Ridd MJ. Monoamine oxidase: From genes to behavior. Annual Review of Neuroscience. 1999;**22**:197-217

[2] Shih JC, Wu JB, Chen K. Transcriptional regulation and multiple functions of MAO genes. Journal of Neural Transmission. 2011;**118**(7):979-986

[3] Meyer JH, Ginovart N, Boovariwala A, Sagrati S, Hussey D, Garcia A, et al. Elevated monoamine oxidase a levels in the brain: An explanation for the monoamine imbalance of major depression. Archives of General Psychiatry. 2006;**63**:1209-1216

[4] Schwartz TL. A neuroscientific update on monoamine oxidase and its inhibitors. CNS Spectr 2013; Suppl 1: 25-32.

[5] Naoi M, Maruyama W, Akao Y, Yi H, Yamaoka Y. Involvement of type a monoamine oxidase in neurodegeneration: Regulation of mitochondrial signaling leading to cell death or neuroprotection. Journal of Neural Transmission. Supplementum. 2006;**71**:67-77

[6] Ou XM, Chen K, Shih JC. Monoamine oxidase a and repressor R1 are involved in apoptotic signaling pathway. Proc Natl Acad Sci USA. 2006;**103**:10923-10928

[7] Brunner HG, Nelen M, Breakefield XO, Ropers HH, van Oost BA. Abnormal behavior associated with a point mutation in the structural gene for monoamine oxidase a. Science. 1993;**262**:578-580

[8] Adeghate E, Parvez H. The effect of diabetes mellitus on the morphology and physiology of monoamine oxidase in the pancreas. Neuro Toxicol. 2004;**25**:167-173

[9] Panagiotidis G, Lindstrom P, Stenstrom A, Lundquist I. Glucose modulation of islet monoamine oxidase activity in lean and obese hyperglycemic mice. Metabolism. 1993;**42**:1398-1404

[10] Bianchi P, Kunduzova O, Masini E, Cambon C, Bani D, et al. Oxidative stress by monoamine oxidase mediates receptor-independent cardiomyocyte apoptosis by serotonin and postischemic myocardial injury. Circulation. 2005;**112**:3297-3305

[11] Kaludercic N, Carpi A, Menabo R, Di Lisa F, Paolocci N. Monoamine oxidases (MAO) in the pathogenesis of heart failure and ischemia/reperfusion injury. Biochimica et Biophysica Acta. 1813;**2011**:1323-1332

[12] Pchejetski D, Kunduzova O, Dayon A, Calise D, Seguelas MH, et al. Oxidative stress-dependent sphingosine kinase-1 inhibition mediates monoamine oxidase A-associated cardiac cell apoptosis. Circulation Research. 2007;**100**:41-49

[13] Cathcart MK, Bhattacharjee A. Monoamine oxidase a (MAO-A): A signature marker of alternatively activated monocytes/macrophages. Inflamm Cell Signal. 2014;**1**:152-159

[14] Trachootham D, Alexandre J, Huang P. Targeting cancer cells by ROS-mediated mechanisms: A radical therapeutic approach? Nature Reviews. Drug Discovery. 2009;**8**:579-591

[15] Rybaczyk LA, Bashaw MJ, Pathak DR, Huang K. An indicator of cancer: Downregulation of monoamine oxidase-a in multiple organs and species. BMC Genomics. 2008;**9**:134

[16] Rhodes DR, Yu J, Shanker K, Deshpande N, Varambally R, et al. Large-scale meta-analysis of cancer microarray data identifies common

transcriptional profiles of neoplastic transformation and progression. Proceedings of the National Academy of Sciences of the United States of America. 2004;**101**:9309-9314

[17] NCBI Gene Expression Omnibus profiles http://www.ncbi.nlm.nih.gov/sites/entrez?db=geo

[18] Peehl DM, Coram M, Khine H, Reese S, Nolley R, et al. The significance of monoamine oxidase-a expression in high grade prostate cancer. The Journal of Urology. 2008;**180**:2206-2211

[19] Wu JB, Shao C, Li X, Li Q, Hu P, et al. Monoamine oxidase a mediates prostate tumorigenesis and cancer metastasis. The Journal of Clinical Investigation. 2014;**124**:2891-2908

[20] Wu JB, Yin L, Shi C, Li Q, Duan P, et al. MAOA-dependent activation of Shh-IL6-RANKL signaling network promotes prostate cancer metastasis by engaging tumor-stromal cell interactions. Cancer Cell. 2017;**31**:368-382

[21] Li J, Yang XM, Wang YH, Feng MX, Liu XJ, et al. Monoamine oxidase a suppresses hepatocellular carcinoma metastasis by inhibiting the adrenergic system and its transactivation of EGFR signaling. Journal of Hepatology. 2014;**60**:1225-1234

[22] Huang L, Frampton G, Rao A, Zhang KS, Chen W, et al. Monoamine oxidase a expression is suppressed in human cholangiocarcinoma via coordinated epigenetic and IL-6-driven events. Laboratory Investigation. 2012;**92**:1451-1460

[23] Dhabal S, Das P, Biswas P, Kumari P, Yakubenko VP, et al. Regulation of monoamine oxidase a (MAO-A) expression, activity and function in IL-13-stimulated monocytes and A549 lung carcinoma cells. The Journal of Biological Chemistry. 2018;**293**:14040-14064

[24] Huang B, Zhou Z, Liu J, Wu X, Li X, et al. The role of monoamine oxidase a in HPV-16 E7-induced epithelial-mesenchymal transition and HIF-1α protein accumulation in non- small cell lung cancer cells. International Journal of Biological Sciences. 2020;**16**:2692-2703

[25] Yamamoto S. Mammalian lipoxygenases: Molecular structures and functions. Biochimica et Biophysica Acta. 1992;**1128**:117-131

[26] Wittwer J, Hersberger M. The two faces of the 15-lipoxygenase in atherosclerosis. Prostaglandins, Leukotrienes, and Essential Fatty Acids. 2007;**77**:67-77

[27] Serhan CN, Chiang N, Van Dyke TE. Resolving inflammation:Dual anti-inflammatory and pro-resolution lipid mediators. Nature Reviews. Immunology. 2008;**8**:349-361

[28] Cathcart MK, Folcik VA. Lipoxygenases and atherosclerosis:Protection versus pathogenesis. Free Radical Biology & Medicine. 2000;**28**:1726-1734

[29] Ku¨hn H, O'Donnell VB. Inflammation and immune regulation by 12/15-lipoxygenases. Progress in Lipid Research 2006; 45: 334-356.

[30] Bhattacharjee A, Shukla M, Yakubenko VP, Mulya A, Kundu S, Cathcart MK. IL- 4 and IL-13 employ discrete signaling pathways for target gene expression in alternatively activated monocytes/macrophages. Free Radical Biology & Medicine. 2013;**54**:1-16

[31] Xu B, Bhattacharjee A, Roy B, Xu HM, Anthony D, et al. Interleukin-13 induction of 15-lipoxygenase gene expression requires p38

mitogen-activated protein kinase mediated serine 727 phosphorylation of Stat1 and Stat3. Molecular and Cellular Biology. 2003;**23**:3918-3928

[32] Gupta V, Khan AA, Sasi BK, Mahapatra NR. Molecular mechanism of monoamine oxidase a gene regulation under inflammation and ischemia-like conditions: Key roles of the transcription factors GATA2, Sp1 and TBP. Journal of Neurochemistry. 2015;**134**:21-38

[33] Bhattacharjee A, Mulya A, Pal S, Roy B, Feldman GM, Cathcart MK. Monocyte 15-lipoxygenase gene expression requires ERK1/2 MAPK activity. Journal of Immunology. 2010;**185**:5211-5224

[34] Bhattacharjee A, Xu B, Frank DA, Feldman GM, Cathcart MK. Monocyte 15- lipoxygenase expression is regulated by a novel cytosolic signaling complex with protein kinase C delta and tyrosine-phosphorylated Stat3. Journal of Immunology. 2006;**177**:3771-3781

[35] True L, Coleman I, Hawley S, Huang CY, Gifford D, et al. A molecular correlate to the Gleason grading system for prostate adenocarcinoma. Proc Natl Acad Sci USA. 2006;**103**:10991-10996

[36] Cao H, Zhang J, Liu H, Wan L, Zhang H, et al. IL-13/STAT6 signaling plays a critical role in the epithelial-mesenchymal transition of colorectal cancer cells. Oncotarget. 2016;7:61183-61198

[37] Lamouille S, Xu J, Derynck R. Molecular mechanisms of epithelial-mesenchymal transition. Nature Reviews. Molecular Cell Biology. 2014;**15**:178-196

[38] Said NA, Williams ED. Growth factors in induction of epithelial-mesenchymal transition and metastasis. Cells, Tissues, Organs. 2011;**193**:85-97

[39] Xiao D, He J. Epithelial mesenchymal transition and lung cancer. Journal of Thoracic Disease. 2010;**2**:154-159

[40] Jiang W, Pang XG, Wang Q, Shen YX, Chen XK, Xi JJ. Prognostic role of twist, slug, and Foxc2 expression in stage I non-small-cell lung cancer after curative resection. Clinical Lung Cancer. 2012;**13**:280-287

[41] Prudkin L, Liu DD, Ozburn NC, Sun M, Behrens C, et al. Epithelial-to-mesenchymal transition in the development and progression of adenocarcinoma and squamous cell carcinoma of the lung. Modern Pathology. 2009;**22**:668-678

[42] Lim J, Thiery JP. Epithelial-mesenchymal transitions: Insights from development. Development. 2012;**139**:3471-3486

[43] Gao D, Joshi N, Choi H, Ryu S, Hahn M, et al. Myeloid progenitor cells in the premetastatic lung promote metastases by inducing mesenchymal to epithelial transition. Cancer Research. 2012;**72**:1384-1394

[44] Kong FF, Qu ZQ, Yuan HH, Wang JY, Zhao M, et al. Overexpression of FOXM1 is associated with EMT and is a predictor of poor prognosis in non-small cell lung cancer. Oncology Reports. 2014;**31**:2660-2668

[45] Liu F, Hu L, Ma Y, Huang B, Xiu Z, et al. Increased expression of monoamine oxidase a isassociated with epithelial to mesenchymal transition andclinicopathological features in non-small cell lung cancer. Oncology Letters. 2018;**15**:3245-3251

[46] Yang X, Zhao D, Li Y, Li Y, Cui W, et al. Potential monoamine oxidase a inhibitor suppressing paclitaxel-resistant non-small cell lung cancer metastasis and growth. Thoracic Cancer. 2020. DOI: 10.1111/1759-7714.13617

[47] Liu Z, Huang L, Luo XJ, Wu L, Li M. MAOA variants and genetic susceptibility to major psychiatric disorders. Molecular Neurobiology. 2016;**53**:4319-4327

[48] Voltas N, Aparicio E, Arija V, Canals J. Association study of monoamine oxidase-a gene promoter polymorphism (MAOA-uVNTR) with self-reported anxiety and other psychopathological symptoms in a community sample of early adolescents. Journal of Anxiety Disorders. 2015;**31**:65-72

[49] Biswas P, Dhabal S, Das P, Das P, Swaroop S, et al. Role of monoamine oxidase a (MAO-A) in cancer progression and metastasis. Cancer Cell & Microenvironment. 2018;**5**:e1623

[50] Zhao H, Flamand V, Peehl DM. Anti-oncogenic and pro-differentiation effects of clorgyline, a monoamine oxidase a inhibitor, on high grade prostate cancer cells. BMC Medical Genomics. 2009;**2**:55

[51] Wu JB, Lin TP, Gallagher JD, Kushal S, Chung LW, et al. Monoamine oxidase a inhibitor-near-infrared dye conjugate reduces prostate tumor growth. Journal of the American Chemical Society. 2015;**137**:2366-2374

[52] Wang K, Luo J, Yeh S, You B, Meng J, et al. The MAO inhibitors phenelzine and clorgyline revert enzalutamide resistance in castration resistant prostate cancer. Nature Communications. 2020;**11**:2689-2703

[53] Lee HT, Choi MR, Doh MS, Jung KH, Chai YG. Effects of the monoamine oxidase inhibitors pargyline and tranylcypromine on cellular proliferation in human prostate cancer cells. Oncol Reports. 2013;**30**:1587-1592

[54] Zarmouh NO, Messeha SS, Mateeva N, Gangapuram M, Flowers K et al. The antiproliferative effects of flavonoid MAO inhibitors on prostate cancer cells. Molecules 2020; 25: 2257-2286.

Chapter 7

Bulk and Nanocatalysts Applications in Advanced Oxidation Processes

Luma Majeed Ahmed

Abstract

Advanced oxidation processes (AOPs) are considered to be vital methods for treating the contaminations produced mainly by the human activations. In present-day, UV light or solar light, bulk and nano- photocatalysts are often used to enhance this technology by creating the highly reactive species such as the hydroxyl radicals. Extreme hydroxyl radical is considered as a key to start the photoreaction. Photoreaction is widely used in treatment of Lab and industrial contaminations, preparation of compounds and produced the renewable energy, so it's classified as green technique. In order to improve the efficiency of this reaction with fabrication the surface of the used photocatalyst such as metal doped, sensitized and produced a composite as bulk catalyst or nano catalyst.

Keywords: nanocatalysts, bulk catalyst, advanced oxidation processes, wastewater treatment, photocatalysis, Fenton reaction, photo-Fenton

1. Introduction

In this section, the advanced Oxidation Processes concepts will be related to use of the bulk and the nano- catalysts as vital materials for easily generating a highly oxidizing species and reactive oxygen species (ROSs) such as in aqueous or alcoholic solution [1]. ROSs are contains three primary kinds: superoxide anion ($O_2^{\bullet-}$), hydrogen peroxide (H_2O_2) and the hydroxyl radical ($HO^{\bullet}$) [2], which produced from reaction of adsorbed oxygen molecule on catalyst's surface with one electron in conductive band under illumination by light as UV, or visible or solar light, this mechanism is useful to reduce the recombination process and increased the life time of hole in valance band [3, 4]. As explained in **Figure 1**.

The ROSs are having the electron configurations as tabled in **Table 1** [5–8].

2. Advance oxidation process applications

In the last few years, several researches have predominated in many universities and research centers on the scientific ventures to mainly treat the contaminations that produced by textile factories [9–11], reduced the degradation of food's

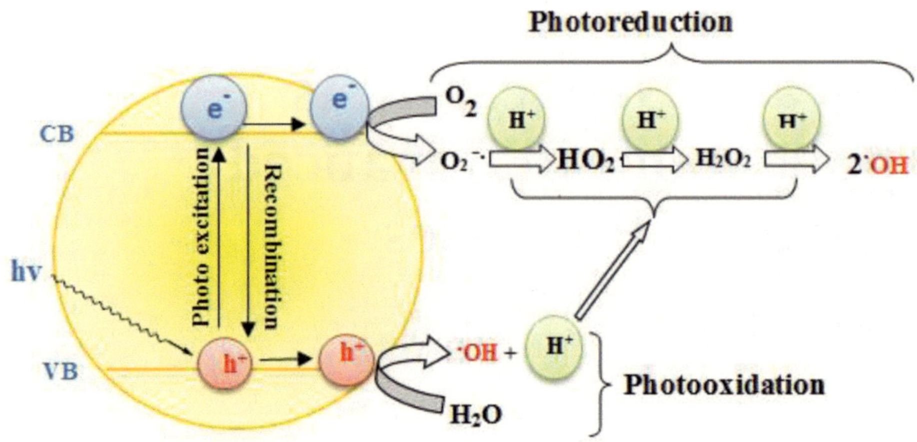

Figure 1.
Essential mechanism for generating the ROSs under illumination of photo-catalyst particles [1].

Oxidation and reactive radical species types	Electronic configurations	Chemical formulas	Oxidation potential V
Oxygen molecule	σ* 2p: — π* 2p: ↑ ↑ π 2p: ↑↓ ↑↓ σ 2p: ↑↓	O_2	1.23
Molecular singlet oxygen	σ* 2p: — π* 2p: ↑↓ — π 2p: ↑↓ ↑↓ σ 2p: ↑↓	$^1O_2^*$	2.42
Superoxide radical anion	σ* 2p: — π* 2p: ↑↓ ↑ π 2p: ↑↓ ↑↓ σ 2p: ↑↓	$O_2^{\bullet-}$	-0.33
Peroxide ion	σ* 2p: — π* 2p: ↑↓ ↑↓ π 2p: ↑↓ ↑↓ σ 2p: ↑↓	O_2	1.78

Table 1.
Electronic configurations and chemical formulas for the ROSs types.

dye [12], decolorization of colored organometallic complexes [13], degradation of toxic cyclic compounds [14] and produced a hydrogen from alcohol as renewable energy [15]. The effective materials for all above mention research are generated the hydroxyl radical in aqueous solution with maximum oxidation power equals to 2.8 V [1]. Based on to the AOPs, the common sources for creation

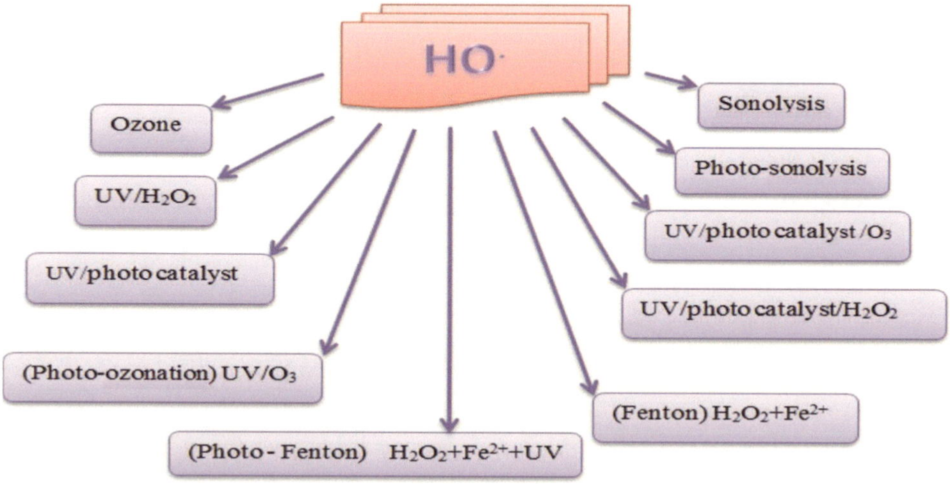

Figure 2.
Schematic diagram of common sources ofOH in advanced oxidation processes.

ofOH in AOPs are illustrated in **Figure 2**, which regards as power to star the dark or photo reactions [1, 16–19].

Fortunately, the benefits of AOPs are more than those of drawbacks. The benefits of AOPs are summarized up as [1, 20] follows to:

1. Create a large number of free radicals species.

2. Have the appropriate potential to depress the hazardous organic pollutants by complete their mineralization and producing CO_2 and H_2O.

3. Reduce the time of dark or photoreaction.

4. Have low economic cost.

Whereas, the drawbacks of AOPs [1, 21] are quenching the reaction rate with increasing the scavenger contains (mostly peroxide ion) and may be generated the undesirable hazardous products that prevented the complete of mineralization process, hence, the altered of pH or using further cost steps may be essentially to treat their problems.

3. Bulk and nano-catalysts

In general, the catalysts may be metal or alloy or semiconductor. Semiconductor is wide used as catalyst and can be element or compound as amorphous or crystalline or rock salt crystal. Because of semiconductors have intermediate properties between metal and insulator, which has given them rescannable electronic and structural properties, hence, semiconductor is classified as a better-known kinds, as mentioned in **Figure 3** [22–24].

The usages of the bulk and nano catalysts are increment with increasing the development of life activations. The catalysts were known for the long time to increase the rate of reaction with decreasing the time of reaction and the activation energy in dark reaction or photoreaction. In order to use the catalyst in

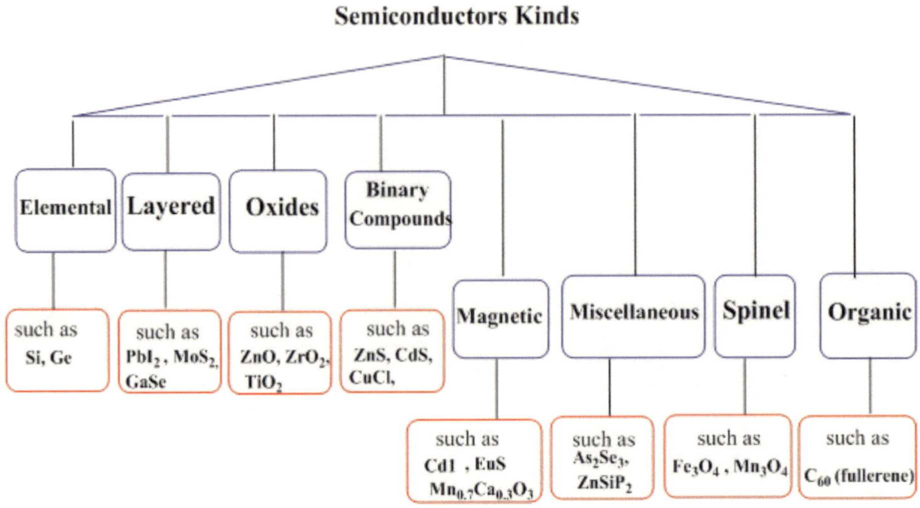

Figure 3.
Better-known kinds of semiconductors.

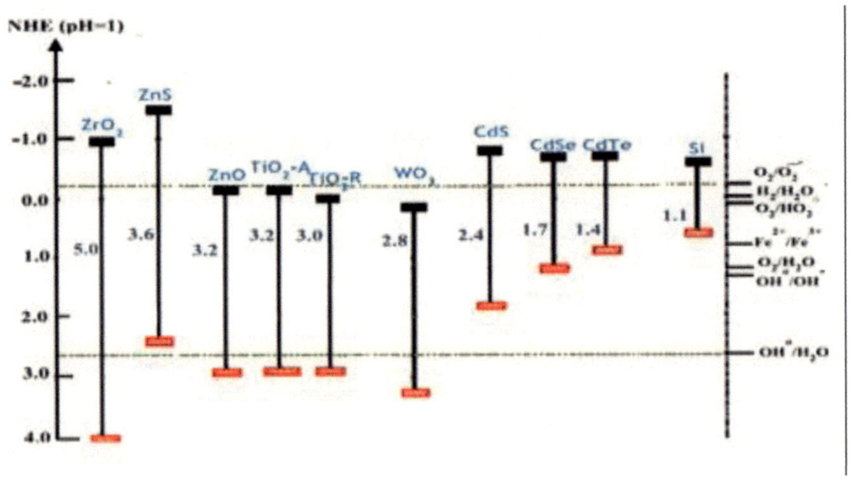

Figure 4.
Band gap energy positions of different photo-semiconductor at pH = 1.

photoreaction as photo catalyst, must have a band gap with raged about 1.1 eV to 5.0 eV [1, 24]. Referring to **Figure 4**, several band gap energy positions of some common photo catalysts can be displayed [1, 25–27].

The mainly problem in bulk and nano catalyst is recombination process, which results in diminishing the efficiency of used photocatalyst by returning the photoelectron from conductive band to valance band and reacting with photohole immediately. The recombination includes four kinds can be followed in **Table 2** and **Figure 5** [1, 28–30].

In order to improve the activity of photocatalysts must depress the recombination with modify their surfaces with three main methods: surface sensitization, metalized photocatalyst surface and coupled for two or more photocatalysts as Composite. The details of these modification methods are mention in **Table 3** and **Figure 6** [40].

Kinds	Other name	Info	Type of photocatalyst
Direct recombination	Band-to- band recombination	In this kind, the transition occurrs as a radiative transition in direct band gap semiconductor. It is created when the Free photo electron in CB drops directly into free photo hole (an unoccupied state) in the VB and associated together. Note **Figure 5(A)**.	ZnO have a direct band gap.
Volume recombination	Centers recombination or Trap-assisted recombination	This case obtains, when defect of semiconductor by impurities that given a new levels (as traps of photoelectron and photohole). It leads to liberate heat as phonon in indirect band gap semiconductor. Note **Figure 5(B)**.	Pure TiO_2 and defect of TiO_2 by metal, which had given an indirect band gap.
Surface recombination	Recombination of an exciton	This case occurs at low temperature, when the traps at or near the surface or interface of the semiconductor, capture the photo electron- hole as exciton. That attitude to dangling bonds caused by the sudden discontinuation of the semi-conductor crystal with energy just below the band gap value. Note **Figure 5(C)**.	It happed in solar cells and light emitting diode (LED) containing shallow levels.
Auger recombination	—	This recombination involves three carriers: Free photo electron, free photo whole recombine, and the emitting the energy as heat or as a photon (non-radiative process). The transition of energy deals with as intra-band transitions, which resulting when either electron elevates in higher levels of conduction band or hole deeper push into the valence band. Note **Figure 5(D)**.	This case can be obtained wit short lifetime when heavy doping defects (like Ag) in direct-gap semiconductors under present sunlight.

Table 2.
The most common recombination types concepts.

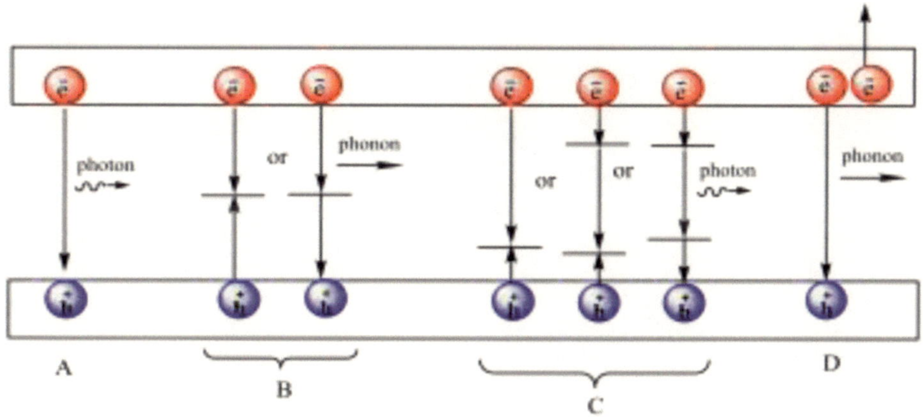

Figure 5.
The schematic diagram of the most common recombination kinds.

Kinds	Info	References
Surface sensitization	This case favors for modified the wide band gap semiconductor by physical or chemical adsorption of colored materials mostly dye. The colored material will absorb the visible or solar light after irradiation, and excite it either singlet or triplet excited state. The excited colored material will inject its electron via the conductive band of semiconductor.	[31–34]
Metalized photocatalyst surface	The metal deposits on the surface of semiconductor must choose with high work functions of metal compared to work function of the metal in semiconductor. The doped metal will act as sink of electron, with create a Schottky barrier. That will increase the lifetime of photo hole. Examples: Pt and Au doped on TiO2, Ag doped on ZnO, Cr and Mn doped on ZnS .	[2, 4, 14, 35–37]
Coupled for two or more photocatalysts as composite	When the energy of the irradiated light is not enough to promote electron from conductive band of the photocatalyst, that attitude to it has a big band gap, hence, can couple it with other semiconductor has a small band gap. This coupling process includes three kinds: type (I), type(II) and type(III). 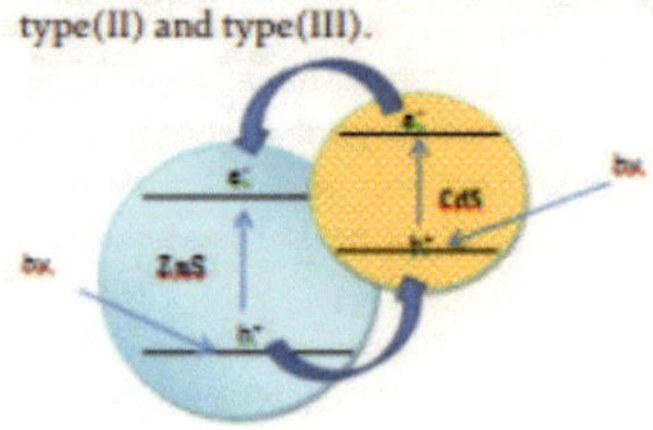 Type I 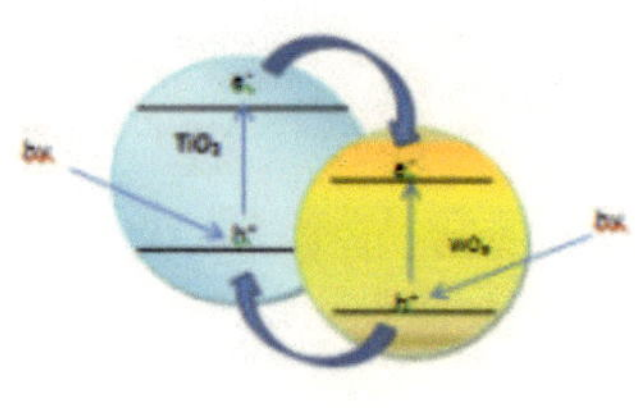 Type II 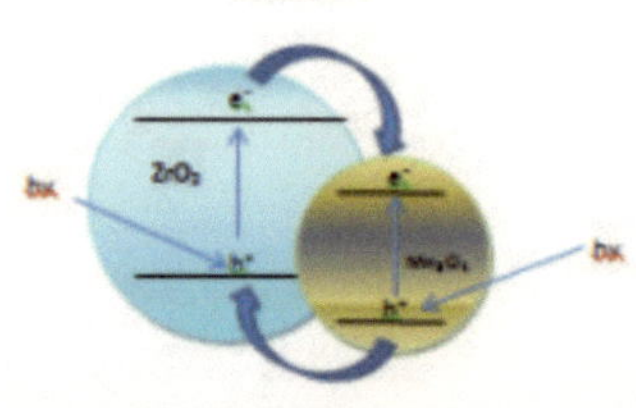 Type III Examples: CdS-ZnS (Type I), WO_3-TiO_2 (Type II). And Mn_3O_4-ZrO_2 (Type III).	[16, 30, 38–39]

Table 3.
The description of the methods for modifying photocatalysts [31–39].

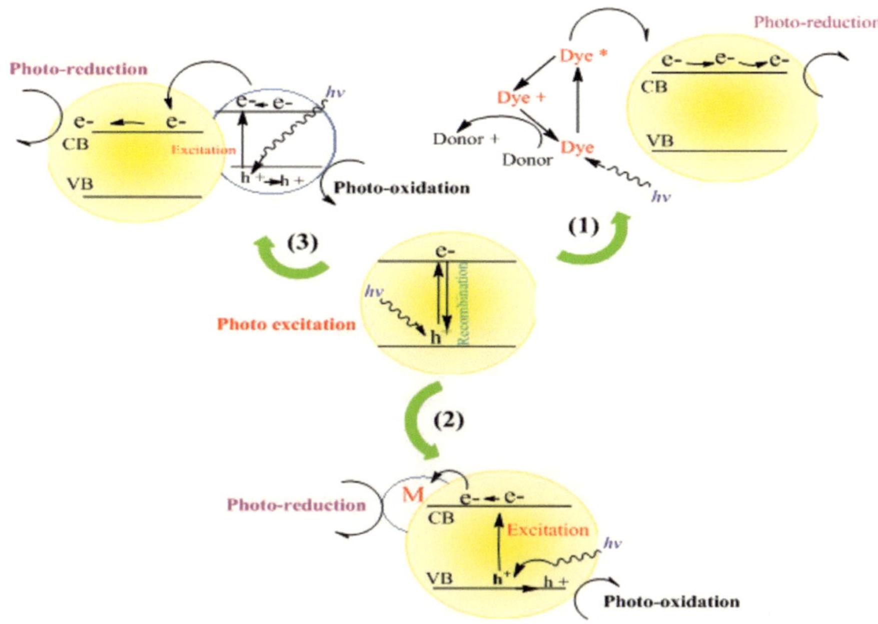

Figure 6.
Schematic diagram for modification of photocatalyst surface [40].

Application field	Type of used AOPs	Efficiency	References
Textile dye Reactive red 2 dye	O_2/UV-A(250 W)/ZnO/ H_2O_2	89.8% (Photodecolorization) (5 mmole/L) of H_2O_2 (T = 25°C), (pH = 10)	[41]
Textile dye direct orange dye	O_2/UV-A(250 W)/ZnO	92.7% (Photodecolorization) (T = 35°C), (pH = 6.68)	[42]
Textile dye reactive yellow 14 dye	O_2/UV-A(250 W)/ZnO	91.41% (Photodecolorization) (T = 38°C), (pH = 6.75)	[43]
Industrial dye Chlorazol black BH dye	O_2/UV-A/ZnO	99.07% (Photodecolorization) (T = 15°C), (pH = 7.63)	[44]
Industrial dye Acid Red 87(Eosin (Eosin Yellow) dye	O_2/UV-A(125 W)/ZnO	74.4.5% (Photodecolorization) (T = 38°C), (pH = 8.6)	[32]
	O_2/UV-A(250 W)/ZnO	98.5% (Photodecolorization) (T = 38°C), (pH = 8.6)	
	O_2/Solar/ZnO	96.5% (Photodecolorization) (T = 42°C), (pH = 8.6)	
Textile dye Dispersive yellow 42 dye	O_2/UV-A(125 W)/ZnO	94.40% (Photodecolorization) (T = 20°C), (pH = 7.7)	[10]
	O_2/UV-A(125 W)/ZnO/Fe^{2+}	60.86% (Photodecolorization) (T = 20°C), (pH = 7.7)	
	O_2/UV-A(125 W)/ZnO/ Fe^{2+}+1% H_2O_2	16.44% (Photodecolorization) (5 x 10^{-4} mole/L) of Fe^{2+} (T = 20°C), (pH = 7.7)	

Application field	Type of used AOPs	Efficiency	References
Drug dye Cobalamine(Vit B12)	O_2/UV-A(250 W)/ZnO	79.33% (Photodecolorization) (T = 30°C), (pH = 6.5)	[19]
	O_2/UV-A(250 W)/ZnO/ $K_2S_2O_8$	88.75% (Photodecolorization) (1 x 10^{-4} mole/L) of $K_2S_2O_8$ (T = 30°C), (pH = 6.5)	
	O_2/UV-A(250 W)/ZnO/ 0.025% H_2O_2	90.80% (Photodecolorization) (T = 30°C), (pH = 6.5)	
	O_2/UV-A(250 W)/ZnO/ $K_2S_2O_8$ + 0.025% H_2O_2	95.85% (Photodecolorization) (1 x 10^{-4} mole/L) of $K_2S_2O_8$ (T = 30°C), (pH = 6.5)	
Food dye Carmoisine (E122) dye	O_2/UV-A(250 W)/ZnO	73.11% (Photodecolorization) (T = 18°C), (pH = 7.55)	[12]
	O_2/UV-A(250 W)/ZnO/ 0.1% H_2O_2	62.58% (Photodecolorization) (T = 18°C), (pH = 7.55)	
	O_2/UV-A(250 W)/ZnO/ Fe^{2+}	36.99% (Photodecolorization) (1 x 10^{-5} mole/L) of Fe^{2+} (T = 18°C), (pH = 7.55)	
Lab materials Co(II) Complex of Schiff Base	O_2/UV-A(250 W)/ZnO	99.11% (Photodecolorization) (T = 38°C), (pH = 7.55)	[13]
Industrial dye Methyl green dye	O_2/UV-A(400 W)/ ZnO NPS	37% (Photodecolorization) (T = 25°C), (pH = 5.4)	[35]
	O_2/UV-A(400 W)/Ag(2%) ZnO NPs	87.37% (Photodecolorization) (T = 25°C), (pH = 5.4)	
Liberated of hydrogen from Methanol as renewable energy	Ar/UV-B(1000 W)/ (0.5 Pt) TiO_2 NPS	8.8% (Photo hydrogen production) (T = 25°C), (pH = 7.3)	[14]
	Ar/UV-B(1000 W)/ (0.5 Au) TiO_2 NPS	4.5% (Photo hydrogen production) (T = 25°C), (pH = 7.3)	
Industrial dye Light Green SF Yellowish (Acid Green 5) Dye	O_2/UV-A(400 W)/ TiO_2	90.2% (Photodecolorization) (T = 20°C), (pH = 7.3)	[45]
	O_2/UV-A(400 W)/ TiO_2 NPS	88.1% (Photodecolorization) (T = 20°C), (pH = 7.3)	

Application field	Type of used AOPs	Efficiency	References
Industrial dye Safranine O Dye	O_2/UV-A(125 W)/ TiO_2 NPS	90.2% (Photodecolorization) (T = 30°C), (pH = 6)	[34]
	O_2/UV-A(125 W)/ TiO_2 NPS/ Fe^{2+}		
	O_2/UV-A(125 W)/ TiO_2 NPS/ Fe^{2+}	85.92% (Photodecolorization) (1 x 10^{-4} mole/L) of Fe^{2+} (T = 30°C), (pH = 6)	
	O_2/UV-A(125 W)/ TiO_2 NPS/ 0.1% H_2O_2	92.73% (Photodecolorization) (T = 30°C), (pH = 6)	
	O_2/UV-A(125 W)/ TiO_2 NPS/ 0.1% H_2O_2+ Fe^{2+}	98.83% (Photodecolorization) (1 x 10^{-4} mole/L) of Fe^{2+} (T = 30°C), (pH = 6)	
Industrial dye Acid Red 87 (Eosin Yellow) dye	O_2/UV-A(250 W)/ TiO_2 NPS	63.58% (Photodecolorization) (T = 25°C), (pH = 6.09)	[16]
	O_2/UV-A(250 W)/ TiO_2 NPS+ H_2O_2	50.44% (Photodecolorization) (1 x 10^{-2} mmole/L) of H_2O_2 (T = 25°C), (pH = 6.09)	
	O_2/UV-A(250 W)/ WO_3 NPS	27.84% (Photodecolorization) (T = 25°C), (pH = 6.09)	
	O_2/UV-A(250 W)/ WO_3 NPS+ H_2O_2	21.54% (Photodecolorization) (1 x 10^{-2} mmole/L) of H_2O_2 (T = 25°C), (pH = 6.09)	
	O_2/UV-A(250 W)/ (0.5) WO_3-TiO_2 nanocomposite	25.11% (Photodecolorization) (T = 25°C), (pH = 6.09)	
	O_2/UV-A(250 W)/ (0.5) WO_3-TiO_2 nanocomposite+ H_2O_2	73.88% (Photodecolorization) (1 x 10^{-2} mmole/L) of H_2O_2 (T = 25°C), (pH = 6.09)	
Industrial dye Methyl green dye	O_2/UV-A(250 W)/ZrO_2	92.31% (Photodecolorization) (T = 30°C), (pH = 5.4)	[46]
	O_2/UV-A(250 W)/ ZrO_2 + Fe^{2+}	39.93% (Photodecolorization) (1 x 10^{-4} mmole/L) of Fe^{2+} (T = 30°C), (pH = 5.4)	
	O_2/UV-A(250 W)/ ZrO_2 + 1.5% H_2O_2	98.78% (Photodecolorization) (T = 30°C), (pH = 5.4)	
	O_2/UV-A(250 W)/ ZrO_2 + $K_2S_2O_8$	74.62% (Photodecolorization) (1 x 10^{-4} mmole/L) of $K_2S_2O_8$ (T = 30°C), (pH = 5.4)	
Lab materials Fe(II)-(4,5-DIAZAFLUOREN-9-ONE 11) COMPLEX	O_2/UV-A(400 W)/ Mn_3O_4	22.64% (Photodecolorization) (T = 15°C), (pH = 4)	[47]
	O_2/UV-A(400 W)/ (1)Mn_3O_4- (4) ZrO_2 nanocomposite	40% (Photodecolorization) (T = 17°C), (pH = 4)	

Application field	Type of used AOPs	Efficiency	References
Textile dye Reactive blue 5 dye	O_2/UV-A(400 W)/ ZnS NPs	59% (Photodecolorization) (T = 15°C), (pH = 6.3)	[36]
	O_2/UV-A(400 W)/ Cr-ZnS NPs	94% (Photodecolorization) (T = 17°C), (pH = 4.1)	
Industrial dye Congo red dye	O_2/UV-A(400 W)/ ZnS NPs	95% (Photodecolorization) (T = 30°C), (pH = 7.5)	[39]
	O_2/UV-A(400 W)/ CdS-ZnS nanocomposite	98% (Photodecolorization) (T = 30°C), (pH = 7.5)	

Table 4.
Some applications of bulk and nano photocatalydts in AOPs, with environment chemistry and green chemistry.

4. Used of bulk or nano catalyst in AOPs

There are many common application of AOPs in environment fields by using the white photocatalyst or its modified such as ZnO, TiO_2 ZrO_2, ZnS, WO_3, CdS and Mn_3O_4. The efficiencies with used these photocatalysts are altered with using AOPs methods. The efficiency of the photoreaction depends mostly on the concentration of colored material, initial pH which affected on the surface of photocatalyst and the temperature. As shown in **Table 4**.

5. Conclusions

This chapter focuses on the source of hydroxyl radical which produces via the advance oxidation process. Indeed, this process interests in the forming the different species, which in the final step generates a hydroxyl radical. The photocatalyst enhances the generating of hydroxyl radicals (2.8 V) in aqueous solution under Uv- light or visible or solar. The photoexitation of photocatalyst leads to jump of electon to conductive band then return to valance band and liberates a hot this process called recombination. It is depressed the efficiency of photoreaction. However, some procedures used to modify the photocatalyst surface.

Acknowledgements

The author wants to thank his family for helping him in carrying out this work.

Author details

Luma Majeed Ahmed
Department of Chemistry, College of Science, University of Kerbala, Kerbala, Iraq

*Address all correspondence to: luma.ahmed@uokerbala.edu.iq

References

[1] Ahmed L, and Hussein F. Roles of Photocatalytic Reactions of Platinized TiO2 Nanoparticales. 1st ed. LAP Lambert Academia Published; 2014. 103 P. ISBN-10: 3659538817.

[2] Collin F. Chemical Basis of Reactive Oxygen Species Reactivity and Involvement in Neurodegenerative Diseases. Int. J. Mol. Sci. 2019; 20, 2407:1-17. DOI: 10.3390/ijms20102407.

[3] Ahmed L, Ivanova I, Hussein F and Bahnemann D. Role of Platinum Deposited on TiO2 in Photocatalytic Methanol Oxidation and Dehydrogenation Reactions. International Journal of Photoenergy. 2014; 1:1-9. DOI: 10.1155/2014/503516.

[4] Ahmed L, Hussen F, Mahdi A. Photocatalytic Dehydrogenation of Aqueous Methanol Solution by Naked and Platinized TiO_2 Nanoparticles. Asian Journal of Chemistry. 2012; 24(12): 5564-5568. DOI:

[5] Krumova K, Cosa G. Chapter 1:Overview of Reactive Oxygen Species . In: Nonell S, Flors C, editors. Singlet Oxygen: Applications in Biosciences and Nanosciences, Volume 1. Uk: Royal social of chemistry; 2016, p. 1-21 DOI: 10.1039/9781782622208-00001

[6] Halliwell B, Gutteridge J. *Free Radicals in Biology and Medicine* , 5th ed. Oxford Oxford: University Press; 2007. 961 P. ISBN 978-0-19-871747-8 (hbk.). ISBN 978-0-19-871748-5 (pbk.)

[7] Sharma S, Ruparelia J, Patel M. A general review on Advanced Oxidation Processes for waste water treatment. Institute of Technology. Nirma University. Ahmed Abad. 2011: 382-481.

[8] Bielski B, Cabelli D, Arudi R, Ross A. Reactivity of HO2/O2 Radicals in Aqueous Solution. J. Phys. Chem. Ref. Data. 1985; 14(4):1041-1100.

[9] Karam F, Saeed N, Al-Yasari A, Ahmed L, Saleh H. Kinetic Study for Reduced the Toxicity of Textile Dyes (Reactive Yellow 14 Dye and Reactive Green Dye) Using UV-A Light/ ZnO System. Egypt. J. Chem. 2020; 63(8): 211-224. DOI: 10.21608/ ejchem.2020.25893.2511.

[10] Alattar R, Saleh H , AL-Hilfi J, Ahmed L. Influence the addition of Fe_2+ and H_2O_2 on removal and decolorization of textile dye (dispersive yellow 42 dye). Egypt. J. Chem., 2020; 63(9): 3191-3201. DOI:10.21608/ EJCHEM.2020.23542.2400.

[11] Kzar K, Mohammed Z, Saeed S, Ahmed L, Kareem D, Hadyi H, Kadhim A. Heterogeneous photo-decolourization of cobaltous phthalocyaninate dye (reactive green dye) catalyzed by ZnO. In AIP Conference Proceedings, 2144(1)020004, AIP Publishing LLC,2019: 020004-01 -020004-10 .

[12] Ahmed L, Jassim M, Mohammed M and Hamza D. Advanced oxidation processes for carmoisine (E122) dye in UVA/ZnO system: Influencing pH, temperature and oxidant agents on dye solution. Journal of Global Pharma Technology. 2018; 10(07): 248-254.

[13] Abass S, Al-Hilfi J, Abbas S and Ahmed L. Preparation, Characterization and Study the Photodecolorization of Mixed-Ligand Binuclear Co(II) Complex of Schiff Base by ZnO. Indonesian Journal of Chemistry. 2020; 20(2):404-412.

[14] Ahmed L, Al-Kaim A, Halbus A, Hussein F. Photocatalytic hydrogen production from aqueous methanol solution over metallized TiO2. International Journal of ChemTech. 2016; 9(10):90-98.

[15] Karam, F, Hussein, F, Baqir S, Halbus A, Dillert R, Bahnemann, D. Photocatalytic degradation of anthracene in closed system reactor. International Journal of Photoenergy. 2014; 1: 1-6.

[16] Jawad T, Ahmed L. Direct Ultrasonic Synthesis of WO_3/TiO_2 Nanocomposites and applying them in the Photodecolorization of Eosin Yellow Dye. Periódico Tchê Química. 2020; 17(34): 621-633.

[17] Stasinakis A. Use of Selected Advanced Oxidation Processes (AOPs) For Wastewater Treatment –A mini review. Global NEST Journal. 2008; 10(3): 376-385.

[18] Marhoon A, Saeed S, Ahmed L. Application of some effects on the Degradation of the aqueous solution of Fuchsine dye by photolysis. Journal of Global Pharma Technology. 2019; 11(9): 76-81.

[19] Ahmed L, Saaed S, Marhoon A. Effect of Oxidation Agents on Photo-Decolorization of Vitamin B12 in the Presence of ZnO/UV-A System. Indones. J. Chem. 2018; 18: 272-278.

[20] Hai F, Yamamoto K, Fukushi K. Hybrid Treatment Systems for Dye Wastewater. Critical Reviews .Environmental Science and Technology. 2007; 37(4): 315-377.

[21] Robinson T, McMullan G, Marchant R, Nigam P. Remediation of dyes in textile effluent: a critical review on current treatment technologies with a proposed alternative. Bioresource Technology. 2001; 77: 247-255.

[22] Yu P, and Cardona M. Fundamentals of Semiconductors Physics and Materials Properties. 4th ed. Germany : Springer-Verlag Berlin Heidelberg; 2010. CH 1.

[23] Yacobi B. Semiconductor Materials- An Introduction to Basic Principles. 1st ed. New York : Kluwer Academic Publishers. 2004. CH 6.

[24] Singh V. Band Gap and Resistivity Measurements of Semiconductor Materials For Thin Films. Journal of Emerging Technologies and Innovative Research (JETIR). 2017; 4(12): 1200-1210.

[25] Material S. Physics Semiconductors and Band Theory, 1st ed. HIGHER: Learning and Teaching Scotland;2011. 4-5 pp.

[26] Carp O, Huisman C and Reller A. Photoinduced reactivity of titanium dioxide, Prog. Solid State Chem. 2004; 32: 33-177.

[27] Kudo A. and Miseki Y. Heterogeneous photocatalyst materials for water splitting. Chem Soc Rev.2009; 38: 253-278.

[28] Kuang W, Tolner H, and Li Q. Cathode-luminescence diagnostics of MgO, MgO:Si, MgO:Sc, and MgCaO . Journal of the SID . 2012; 20(1): 63- 69S.

[29] Pierret R. Advanced semiconductor Fundemantals. 1st ed. vol. VI, Modular series on solid state devices, New Jersey : Pearson Education, Inc. Upper Saddle River;2002. 07458 p. CH5.

[30] Hayawi M. Preparation, Characterization of Spinel Mn3O4/ ZrO2 Composite and Application on Colored Material[MSC thesis]. Karbala: Kerbala University:2020.

[31] Giwa A, Nkeonye P., Bello K., Kolawole G. and Campos A. Solar Photocatalytic Degradation of Reactive Yellow 81 and Reactive Violet 1 in Aqueous Solution Containing Semiconductor Oxides. International Journal of Applied Science and Technology. 2012; 2(4): 90-105.

[32] Ahmed L. Photo-decolorization kinetics of acid red 87 dye in ZnO suspension under different types of UV-A light. Asian J. Chem. 2018; 30(9): 2134-2140.

[33] Pare B, Singh P and Jonnalagadda S, Visible Light-drive Photocatalytic Degradation and Minieralization of neutral Red dye in a sulurry Photoreactor. Indian Journal Chemistry Technoogy.2010; 17: 391-395.

[34] Jasim K, and Ahmed L. TiO2 Nanoparticles Sensitized by Safranine O Dye using UV-A Light System. In IOP Conference Series. Materials Science and Engineering. 2019; 571(012064):1-9.

[35] Fadhil F, Ahmed L, and Mohammed A. Effect of silver doping on structural and photocatalytic circumstances of ZnO nanoparticles. Iraqi Journal of Nanotechnolog, Synthesis and Application. 2020; 1: 13-20.

[36] Mahammed B, and Ahmed L. Enhanced Photocatalytic Properties of Pure and Cr-Modified ZnS Powders Synthesized by Precipitation Method. Journal of Geoscience and Environment Protection. 2017; 5: 101-111.

[37] Mohammed B, and Ahmed L. Improvement the Photo Catalytic Properties of ZnS nanoparticle with Loaded Manganese and Chromium by Co-Precipitation Method. Journal of Global Pharma Technology. 2018; 10(7):129-138.

[38] Jawad T. Synthesis and Characterization of Nano-Composite WO3/ TiO2 by Using Ultrasonic and it is Application on Photodecolorization of Eosin Yellow Dye [MSC thesis]. Karbala: Kerbala University: 2020.

[39] Fakhri F, and Ahmed L. Incorporation CdS with ZnS as nanocomposite and Using in Photo-Decolorization of Congo Red Dye. Indones. J. Chem.2019; 19(4):936-943.

[40] Fakhri F. Preparation, Characterization of ZnS/CdS Composites nano particles and using in photocatalytic-decolorization of Congo red dye[MSC thesis]. Karbala: Kerbala University: 2019.

[41] Mashkour M, Al-Kaim A, Ahmed L, and Hussein F. Zinc Oxide Assisted Photocatalytic Decolourization of Reactive Red 2 Dye. Int. J. Chem. Sci. 2011; 9(3): 969-979.

[42] Zuafuani S, and Ahmed L. Photocatalytic Decolourization of Direct Orange Dye by Zinc Oxide under UV Irradiation. Int. J. Chem. Sci.2015; 13(1):187-196.

[43] Ahmed L, Tawfeeq F, Abed Al-Ameer F, Abed Al-Hussein K. and Athaab A. Photo-Degradation of Reactive Yellow 14 Dye (A Textile Dye) Employing ZnO as Photocatalyst. Journal of Geoscience and Environment Protection.2016; 4: 34-44.

[44] Abbas S, Hassan Z, and Ahmed L. Influencing the Artificial UV-A light on decolorization of Chlorazol black BH Dye via using bulk ZnO Suspensions. In Journal of Physics: Conference Series. IOP Publishing. 052050. 2019; 1294(5):1-8.

[45] Eesa M, Juda A, and Ahmed L. Kinetic and thermodynamic study of the photocatalytic decolourization of light green SF yellowish (acid green 5) dye using commercial bulk Titania and commercial Nanotitania. Int. J. Sci. Res.2016; 5(11): 1495-1500.

[46] Hussein Z, Abbas S, and Ahmed, L. UV-A activated ZrO2 via photodecolorization of methyl green dye. In IOP Conference Series: Materials

Science and Engineering. 012132, IOP Publishing. 2018; 454(1): 1-11.

[47] Hayawi M, Kareem M, and Ahmed L. Synthesis of Spinel Mn3O4 and Spinel Mn3O4/ZrO2 Nanocomposites and Using Them in Photo-Catalytic Decolorization of Fe(II)-(4,5-Diazafluoren-9-One 11) Complex . Periódico Tchê Química. 2020; 17(34): 689-699.

Chapter 8

Oxidoreductases: Significance for Humans and Microorganism

Hussein Mahdi Kareem

Abstract

Oxidoreductases consist of a large class of enzymes catalyzing the transfer of electrons from an electron donor (reductant) to an electron acceptor (oxidant) molecule. Since so many chemical and biochemical transformations comprise oxidation/reduction processes, it has long been an important goal in biotechnology to develop practical biocatalytic applications of oxidoreductases. During the past few years, significant breakthrough has been made in the development of oxidoreductase-based diagnostic tests and improved biosensors, and the design of innovative systems for the regeneration of essential coenzymes. Research on the construction of bioreactors for pollutants biodegradation and biomass processing, and the development of oxidoreductase-based approaches for synthesis of polymers and functionalized organic substrates have made great progress. Proper names of oxidoreductases are in a form of "donor:acceptor oxidoreductase"; while in most cases "donor dehydrogenase" is much more common. Common names also sometimes appeared as "acceptor reductase", such as NAD+ reductase. "Donor oxidase" is a special case when O_2 serves as the acceptor. In biochemical reactions, the redox reactions are sometimes more difficult to observe, such as this reaction from glycolysis: Pi + glyceraldehyde-3-phosphate + $NAD^+ \rightarrow$ NADH + H^+ + 1,3-bisphosphoglycerate, where NAD+ is the oxidant (electron acceptor), and glyceraldehyde-3-phosphate functions as reductant (electron donor).

Keywords: oxidoreductases, important of enzyme, application medical of this enzyme

1. Introduction

1.1 Enzymes

Are biotic chemical agents that rise the amount of biochemical reaction by depressing of activate energy. The particles convoluted in the enzyme intermediated responses is identified as substrate and the outcome of the reactions or produce are termed products. In general, the chemical structure of greatest for more enzymes is protein and hardly ever of other type e.g., Ribonucleic acid (RNA). The enzyme is too special on the way to their substrates of whom they re-join and thereby the reaction will also be so specific. At times the enzymes requests the turnout of a un protein part called coenzyme, if was a vitamin derivative Organic

 IntechOpen

complex or cofactor, if was a metal- ion for obtain the reactions. And for this, entire enzymes might be named a **holoenzyme**, the portion of protein by means of apoenzyme and the nonprotein basic a prosthetical collection.

2. Enzymes oxido-reductases

Oxido-reductases are a great collection of enzyme is existing of differential area in natural lifecycle such as microorganisms, plant and animals. The enzymes commission EC numbers taxonomy of enzymes. They are categorized by way of EC 1. It is include approximately one third of the enzyme actions that are recorded in BR aunschweig Enzyme List (Selles vidal *et al*, 2018). This enzyme stimulate (give-and-take) of electron among the (donor and acceptor) molecule, reaction comprising electrons transferal, protons, Hydrogen extractive, Hydride transfer, Oxygen insert, also extra significant stages [1, 2]. Generally, two in half reaction such as some oxidative and one reduction occurring and at smallest two substrate such as one reduces and one oxidize is activate and convert [3]. Oxidoreductases comprise of a great categorize of enzyme catalyze the transmission of electron from an electrons donor (reduction) to an electron acceptor (oxidation) molecules, general take NADP nicotinamide- adenine- dinucleotide phosphate or NAD nicotinamide –adenine- dinucleotide as cofactor (**Figure 1**) [4]. Then so various biochemical conversions include oxidant –reluctant methods, it has more been a significant aim in biotechnological to progress applied bio- catalytic uses of oxido-reductases. Through the past little years, significantly discovery has been through in the improvement of oxido-reductase-based diagnosis checks than developed bio-sensors and the plan of new system into the renewal of necessary co-enzymes. Study on the structure of bioreactor for contaminants biodegrade and Biomass treating, and the improvement of oxido-reductase-Based styles into production of polymer and functional Organic substrate have prepared grates progresses. Correct name of oxido-reductases is of donor-acceptor oxido-reductase. However in greatest case donor- dehydrogenase is much more public. Public name also at times appeared as (acceptors –reductase) for example NAD + reductase (donor –oxidative) is a specific example when O_2 render as acceptors. He catalyzed reaction are like to the reaction in **Figure 1**- A the reduction and B is oxidative. In active bio-chemical reaction, the reduction reaction are at times extra difficult to detect, example reactions glycolysis (Pi + glycer aldehyde$^-$3$^-$phosphate + NAD → NADH + H + 1,3-Bisphospho glycerate. NAD+ is the oxidant (electron - acceptor), and glyceraldehyde-3-phosphate function as reduction t (electrons -donors).

3. Classification of oxido-reductases

Oxidoreductases may be categorized accord to the arrangement or structure of three dimensions building, that is extremely instructive aimed at the identify of structure functions correlation, enzymes development, function genomic, and

$$A_{red} + B_{ox} \longrightarrow A_{ox} + B_{red}$$

Figure 1.
Oxidation-reduction.

silicon new enzyme detection, for use, Oxido-reductases also may be categorized accord to their name stimulation and or co-enzyme-dependence.

Hydroxylases, oxygenases, peroxidases and reductases (**Figure 2**) [4, 5]. The molecule Oxygen actions as receptor of Hydrogen or electron. Their enzymes called **oxidases** is convoluted. But this enzyme was dehydrogenase, the outcome of which is confirmed by a hydrogen transmission of an accepter r molecule that contains either/or nicotinamide adenines-adenine-dinucleotide phosphate NAD+/NADP+ or a flavic co-enzyme [6].

Peroxydases catalyze the reduction of the addition of hydroxyl to substrates. Oxygenases integrate oxygen into the organic substrates of molecular oxygen.

Reductases stimulate reduce reaction, and in more cases they action similar oxidases. Oxidoreductases accomplish essential role in together Aerobic metabolism and Anaerobic mechanism. They have an extensive variety of substrates, together Organic (alcohol, amine and ketone) and inorganics (some anions like sulfite and some types metal like (Mercury). This enzymes has many reductive -active centers for performance many physiologically functions [7]. This centers safe via the poly peptide backbone of Oxido-reductases as they are very variable in environment. Polypeptides basis are of the enzymes as well supports in Selectivity, reactivity, redox potential, Stability and inhibit resistance. This Public reductive centers comprise amino acid excesses such as (tyrosine-cysteine), metals ions or complex Examples of these are the co-enzymes (c., mo, fe-s), pterion, and pyro-loquinolin (Pquq), for example (cu, mo, fe, fe-s group), and flavin mononucleotide (FMN) (**Figure 2**).

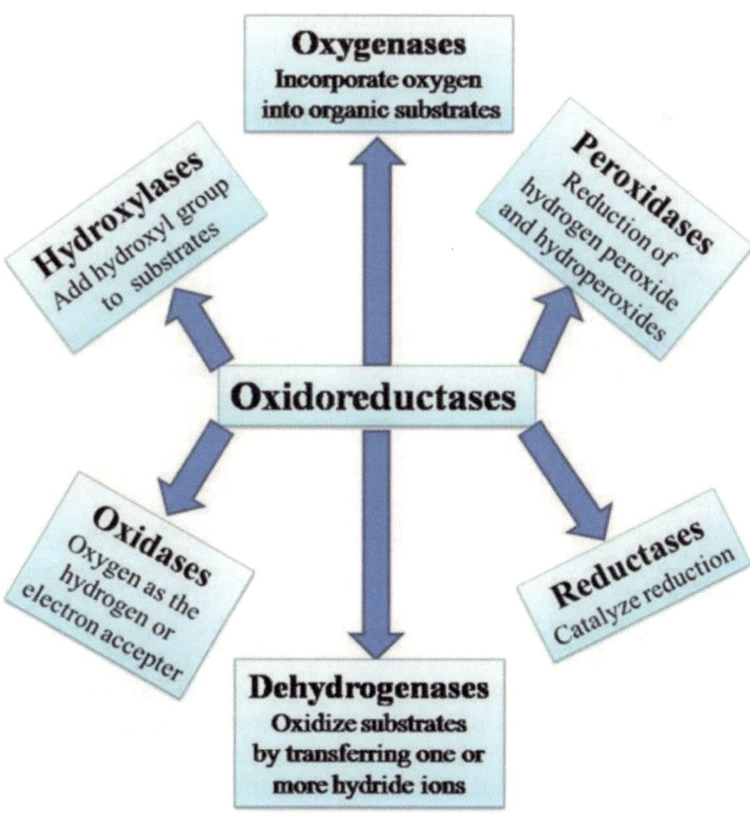

Figure 2.
Classification of oxidoreductases.

4. Applications of oxidoreductases

Since several chemical and biochemical conversions include methods for oxidation reduction, it was attractive, albeit somewhat elusive, to develop developed bio-catalytic uses of oxidation enzymes since the early years of biomedical technologies [8]. Application envision for these enzymes have involved a symmetric oxy functionalization of steroid and other pharmaceutical, production and alteration of polymer, oxidation degrade of contaminants, oxy functionalization of hydrocarbons, and the structure of biosensors for a diversity of analytical and clinical application. Oxidoreductase created catalysis turns well by way of the improvement in greatly effective, maintainable, and medium-friendly industry then they are recyclable, exact in natural surroundings, and energy save. This enzymatic system can include diverse co-factors like Simetric steroid and other pharmaceutical oxygen functions, polymer synthesis and modification, oxidative degradation of pollutants, hydrocarbon Oxyfunction, and a biosensor structure were included in the application of those enzymes for a range of analytical and clinical applications [9].

4.1 Carbohydrates application

Particle carbohydrates can be employed as a renewable resources and cheap rare material, forerunner, building block, or addition for numerous industrial produces. Once, beneficial Organic acid like lactic acid takes been produce from sugar by complete cellular fermentation methods [10]. By the Oxidoreductases uses enzymes, Particle sugar use in everyday our life like particles (glucose, sucrose) can be altered into new beneficial products. Also Particle D-glucose was modified by enzyme glucose Oxidase to Type-glucosone [11]. The cheese processing industry has produced lactose by way of by-products, which has been renewed to lactobionic acid by enzyme lactose oxidase) [12]. Also the lactobionic acid is employed as a worthy diet addition, chelators, acid, and a polymers forerunner [13].

4.2 Conversion of biomass

Conservative dealignment of the pulp is based on a single chlorine or chemical oxidant based on oxygen. While very active, these agents can cause serious problems in the disposal of products or damage to cellulose fiber. Enzymatic delignifying devices are appealing alternatives [14]. Laccase- peroxidase- and other oxidoreductases share in the natural delignification by lignolytic white-rot fungi. Numerous laccases have been shown capable of degrading together natural and artificial lignin (Balakshin *et al*, 2001). They oxidation by direct the phenolic elements of lignin's in the existence of a correct reduction reactions pander, indirect, the hetero geneous Phenolic and non-phenolic chiefly methoxy benzene component. The product, radical can be made in lignin's, which could leads to aliphatic or aromatic C–C connection split and de polymerization. Enzymes Lignins peroxidase is also a strong DE lignifying factor. Its high valent Oxo - Ferryl types can extract electron or proton from the non-phenolic part of structure lignins, therefore producing radical that split the Heterogeneous polymers. Similar enzymes lignin peroxidase, enzymes Mn peroxidase is also working by White Rot Fungus to destroy Lignins. Enzymes Mn peroxidase is of specific importance, for the reason that its oxidation agent. Enzyme Mn peroxidase [III] may stabilize by lesser chelator such as oxalate C_2O_4 (2−) and diffusion on the places in lignins normal impossible in enzyme [15].

4.3 Technologies for textiles

Potential for the use of oxidoreductase in textile manufacture consists of cotton fiber bleaching, dyeing and waste management. The enzyme whitening process of cotton is explained in a recent study [16]. Significant result of lacquer application to the bleaching of the cotton I observed in the peroxide mix. Potential benefits are chemical, energy and saving water Laccase-catalyzing textile dye bleaching is advantageous for the finish of the cotton fabric [17].

4.4 Technologies for food

The essential components of several diets and beverages include many oxidoreductase substrates, including carbohydrates, unsaturated fatties, phenolics, and thiol. The alteration of oxidoreductase may lead to new functionality, quality development, or cost reduction [18]. Often O_2, because of excessive oxidative, is useful in the consistency or storage of food drinks. Oxidase can be used as O_2-scavengers for enhanced food packaging [19]. The promotion of glucose oxidase for bread Makin uses. Addition the enzymes to dough can leads to several chemical physical variations comprising cross-link of protein albumin protein, globulin, and to reduced amount, glutenins [20]. Therefore the paste demonstrations improved viscoelasticity - rheological properties, and the baked baking has better fragment, greater volume, or extra features. The influence is like cause by molecule H_2O_2 made by the enzymes. But, the actions of this enzymes is not higher to that induce oxidation additive like Bromic acid anion, BrO^{-3} and azodicarbonamide. For bread making applications, glucose oxidase has been commercialized. Addition of the enzyme to dough can lead to various physicochemical changes including cross-linking of wheat albumin, globulin, and to some extent, glutenin chemical [21]. So, they are essential to detect or improve other enzyme Carbohydrate oxidases for this enforcement. The lipoxy genase enzymes are a favorable nominee for the sowbread application [22]. The effect of paste strengthening and bread whitening can be achieved with enzymes by modification and emulsifying properties of endogenous fatty acid saturation lipids and the formation of oxide peroxide. But adding enzymes to a certain food may cause the endogenous antioxidant to lack or deplete.

4.5 Bioconversion, biocontrol, and environmental use

Bioconversion of extensively use insecticides, herbicide and many agro chemicals is a significant importance in technological advance the social order, and peroxidase enzyme have great potential for like applications. Mention researcher [23] the ability of Phanerochaetaceae Onygenaceae, Basidiomycota genus Trametes, Tinea versicolor, Coriolopsis gallica and family of fungi Pleurotaceae grow in a nitrogen- contain amount lower of mineral culture media which degradation PCBs was compare, then separate amount of PCBs extracted from these fungal culture media for period four weeks were 25, 50, 41, and O %, respective. Enzymes examines established that both in elevation and comparatively firm activities of all enzymes following: Mn dependent peroxidase, Mn independent peroxidase, lignin peroxidase and lactase described efficacious degradation. In linked works, lactase from Pityriasis versicolor was presented to be qualified for in vitro oxidative of poly cyclic arene hydrocarbons with construction of the congruous Quinone as oxidative produces [24] Amazingly, adding of the pander l-hydroxybenzotriazole for enzyme response solution helped the reactions to such an extent that polycyclic aromatic hydrocarbon (PAH), Fluorenes, solid polycyclic aromatic hydrocarbon, Benzopyrene $C_{20}H_{12}$, and perilene were almost complete remove from the solution.

4.6 Medicine and other synthetic enforcement

The enzyme oxidoreductases are essential in medical combination. For example enzyme Laccases can be employed to produce a great amount of compound medical mediators, like Triazolobenzodiazepine, Cycloalkyl Thiadiazoles, (Cephalosporin β-lactam antibiotics), vincaleukoblastine, Penicillin X methyl ester [25, 26]. The enzymes benzenediol: oxygen oxidoreductases; EC 1.10.3.2 may be application to produce numerous practical Organic combinations include polymer of similar electric optical mechanical characteristics, flavor agent, texture dyes, structure cosmetic pigment, and pesticide [27]. By use of Oxidoreductases can leads for improvement of modern industry artificial techniques. Such as Baeyer-Villiger mono oxygenase can stimulate beneficial expansions of ring reactions by transformation a cyclic ketone to the congruent lactone [28]. Macrophomic acid production enzymes can stimulate Diels alder reactions [29]. At times after the Oxido-reductases performances on its substrates, it can induce a second response with parts of the substrates that lead to modern types of bio catalysis [30]. The use enzyme oxido-reductases we can stimulate reaction that are not simply favorable Such as chloroperoxidase and Cytochrome P450 enzyme can functionalizing indeclinable hydro carbons by hydroxylation [31]. Enzyme enone reductase can Hydrogenation unsaturated bond to change component ketone to hydrocarbons [32]. Old yellow enzyme gained from type of fungi that called Yeast that contain FMN enzyme, can stimulate the reduction by NADPH of the Olefinic (>C¼C<) not carbonyl >C¼O the site of 2-Cyclohexen- [33]. Application oxidoreductases can leads to different industry produce methods. Such as Baeyer-Villiger mono oxygenase can stimulate A valuable ring-expanding reaction by altering a cyclic ketone to a corresponding lactone [28]. Sulfoxidation of alkyl aryl sulfides, nitroso-and-hydroxylamino-compounds N-oxidation, or styrene epoxidation can be done by horseradish peroxidase. Enone reductase can hydrogenate unsaturated bonds to convert ketones to hydrocarbons [32].

Author details

Hussein Mahdi Kareem
Institute of Technical Diwaniyah, Al-Furat Al-Awsat Technical University, Iraq

*Address all correspondence to: mahdihussein73@gmail.com; dw.hus10@atu.edu.iq

References

[1] Husain Q (2017) High yield immobilization and stabilization of oxidoreductases using magnetic nanosupports and their potential applications: an update. Current Catal 6(3):168-187

[2] Toone EJ (2010) Advances in enzymology and related areas of molecular biology, Protein evolution, vol 75. Wiley, Hoboken, NJ

[3] Laskar AA, Alam MF, Younus H (2017) In vitro activity and stability of pure human salivary aldehyde dehydrogenase. Int J Biol Macromol 96:798-806

[4] Nicholas C, Lewis S (1999) Fundamentals of enzymology: the cell and molecular biology of catalytic proteins. Oxford University Press, Oxford

[5] Husain, M. F. Ullah (2019). Biocatalysis Enzymatic Basics and Applications , Springer Nature Switzerland https://doi.org/10.1007/978-3-030-25023-2

[6] Alam MF, Laskar AA, Choudhary HH, Younus H (2016) Human salivary aldehyde dehydrogenase: purification, kinetic characterization and effect of ethanol, hydrogen peroxide and sodium dodecyl sulphate on the activity of the enzyme. Cell Biochem Biophys 74:307-315

[7] Webb EC (1992) Enzyme nomenclature 1992. Recommendations of the Nomenclature Committee of the International Union of Biochemistry and Molecular Biology on the nomenclature and classification of enzymes. Academic, San Diego, CA

[8] May SW. (1999). Applications of oxidoreductases. Current Opinion in Biotechnology, 10(4), 370-375

[9] May SW, Padgette SR (1983) Oxidoreductase enzymes in biotechnology: current status and future potential. Nat Biotech 1:677-686

[10] Ghaffar T, Irshad M, Anwar Z, Aqil T, Zulifqar Z, Tariq A, Kamran M, Ehsan N, Mehmood S (2014) Recent trends in lactic acid biotechnology: a brief review on production to purification. J Radiat Res Appl Sci 7:222-229

[11] Karmali A, Coelho J (2011) Bioconversion of D-glucose into D-glucosone by glucose 2-oxidase from Coriolus versicolor at moderate pressures. Appl Biochem Biotechnol 163:906-917

[12] Koka R, Mehnert D, Fritsch R, Steffan W, Habermeier P, Bradbury A, Wolfschoon-Pombo A, Rose M (2004) Process for manufacturing cheeses and other dairy products and products thereof. In:Google Patents

[13] Gutiérrez LF, Hamoudi S, Belkacemi K (2012) Lactobionic acid: a high value-added lactose derivative for food and pharmaceutical applications. Int Dairy J 26:103-111

[14] Call H. PCT patent WO2003061550-A2 (2003)

[15] Hofrichter M. (2002). Enzyme Microb Technol 30, 454-46

[16] Tzanov T, Basto C, Gübitz GM, and Cavaco-Paulo A. (2003). Macromol Mater Eng. 288, 807-810

[17] Xu F. (1999). in The Encyclopedia of Bioprocessing Technology: Fermentation, Biocatalysis, and Bioseparation, eds. John Wiley & Sons, New York Flickinger MC, and Drew SW. 1545-1554

[18] Minussi RC, Pastore GM, and Durán N. (2002)., Trends Food Sci Technol 13, 205-216

[19] Andersson M, Andersson T, Adlercreutz P, Nielsen T, and Hörnsten EG. (2002). Biotechnol Bioeng 79, 37-42

[20] Rasiah IA, Sutton KH, Low FL, Lin HM, Gerrard JA (2005) Crosslinking of wheat dough proteins by glucose oxidase and the resulting effects on bread and croissants. Food Chem 89:325-332

[21] Kohajdová Z, Karovičová J, Schmidt Š (2009) Significance of emulsifiers and hydrocolloids in bakery industry. Acta Chimica Slovaca 2:46-61

[22] Casey R, West SI, Hardy D, Robinson DS, Wu Z, Hughes RK (1999) New frontiers in food enzymology: recombinant lipoxygenases. Trends Food Sci Technol 10:297-302

[23] Novotny C, Vyas BRM, Erbanova P, Kubatova A, Sasek V (1997). Removal of PCBs by various white rot fungi in liquid cultures. Folia Microbial, 42:136-1 40.

[24] Majcherczyk A, Johannes C, Huettermann A (1998). Oxidation of polycyclic aromatic hydrocarbons (PAH) by lactase of Trametes versicolor. Enzyme Microb Technol, 22:335-341.

[25] Mikolasch A, Niedermeyer THJ, Lalk M, Witt S, Seefeldt S, Hammer E, Schauer F, Gesell Salazar M, Hessel S, Jülich WD (2007) Novel cephalosporins synthesized by amination of 2, 5-dihydroxybenzoic acid derivatives using fungal laccases II. Chem Pharma bulletin 55:412-416

[26] Sagui F, Chirivì C, Fontana G, Nicotra S, Passarella D, Riva S, Danieli B (2009) Laccase-catalyzed coupling of catharanthine and vindoline: an efficient approach to the bisindole alkaloid anhydrovinblastine. Tetrahedron 65:312-317

[27] Ose T, Watanabe K, Mie T, Honma M, Watanabe H, Yao M, Oikawa H, Tanaka I (2003) Insight into a natural Diels-Alder reaction from the structure of macrophomate synthase. Nature 422:185-189

[28] Alphand V, Carrea G, Wohlgemuth R, Furstoss R, Woodley JM (2003) Towards large-scale synthetic applications of Baeyer-Villiger monooxygenases. Trends Biotechnol 21:318-323

[29] Sedmera P, Halada P, Peterbauer C, Volc J (2004a) A new enzyme catalysis: 3,4-dioxidation of some aryl β-d-glycopyranosides by fungal pyranose dehydrogenase. Tetrahedron Lett 45:8677-8680

[30] Kimoto N, Yamamoto H (2004). Novel enone reductases isolated from Kluyveromyces lactis, methods for producing same, and methods for selectively reducing a carbon-carbon double bond of an Alpha, Beta-unsaturated ketone using the reductases. In: Google Patents

[31] Bell SG, Orton E, Boyd H, Stevenson JA, Riddle A, Campbell S, Wong LL (2003) Engineering cytochrome P450cam into an alkane hydroxylase. Dalton Transac 11:2133-2140

[32] Colonna S, Pironti V, Carrea G, Pasta P, Zambianchi F (2004a) Oxidation of secondary amines by molecular oxygen and cyclohexanone monooxygenase. Tetrahedron 60:569-575

[33] Massey V (2000) The chemical and biological versatility of riboflavin. Biochem Soc Trans 28:283-296

Chapter 9

Neurodegeneration: Diagnosis, Prevention, and Therapy

Mrinal K. Poddar, Apala Chakraborty and Soumyabrata Banerjee

Abstract

Neurodegenerative disorders (NDDs) are a broad range of pathological conditions which target the neurons, creating problems in movements and mental functions. The NDDs have drawn a lot of attention among the diseases because of its complexity in causes and symptoms, lack of proper effective treatment(s), no report of irreversibility, and poor impact on social and financial aspects. Individual's vulnerability towards the stress-related biochemical alterations including increase in oxidase enzymes' activities and generation of free radicals, abnormal protein dynamics, mitochondrial dysfunctions, and neuroinflammation often lead to degeneration of neuronal cells. Some advanced techniques are now able to detect the development and progression of different NDDs' complications. The current focus of research on NDDs is to establish convenient therapeutic strategies by targeting different aspects including upliftment of cellular defense mechanisms, especially oxidoreductases as a protective tool. This chapter focused on those updated information on the development, diagnosis, prevention, and therapeutic strategies of NDDs.

Keywords: neurodegenerative disorders, proteinopathies, oxidoreductase, neuroimaging, brain mapping, neurotrophic factors, neuroinflammations, epigenetic modulations

1. Introduction

Neurodegeneration refers to a progressive structural and functional loss of neurons causing heterogeneous clinical and pathological expressions followed by deterioration of functional anatomy [1]. This progressive neuronal cell death often leads to various neurodegenerative disorders (NDDs) such as Parkinson's disease (PD), Huntington's disease (HD), Alzheimer's disease (AD), amyotrophic lateral sclerosis (ALS), brain trauma (BT), prion disease (PrD), progressive supranuclear palsy (PSP), and spinocerebellar ataxias (SCA), etc., which can be differentiated based on their different pathological mechanistic pathways. It includes associated neuropathology, disease based anatomical vulnerability, and aggregation of some major selective proteins during disease conditions [2]. In the last few decades, several approaches have been taken to understand the mechanisms of neuronal cell death [3]. The oxidative and nitrosative stress due to the overproduction of reactive oxygen species (ROS) and reactive nitrogen species (RNS) with the deterioration of cellular antioxidant defense systems are found to be the major reasons behind this neuronal cell damage which might further lead to NDDs [4]. In this context, it is obvious to mention that the oxidoreductase enzymes which are responsible to increase the oxidant level in the cellular microenvironment are one of the major culprits of these sophisticated diseases [4–6]. These pathways and mechanisms of these biochemical processes leading towards the cell deaths are found to be different for various neurodegenerative diseases as observed by their symptoms and exacerbations [4–6]. The common neuropathological hallmarks of such diseases are (a) stress-induced generation of free radicals (b) abnormal protein dynamics, their degradation, and aggregation (c) mitochondrial dysfunctions and (d) neuroinflammation [2] (**Figure 1**). Advanced immunohistochemical and biochemical methods are now able to identify the specific protein abnormalities, related to each of the classes of NDDs [7]. These proteins mostly follow the brain region-specific sequential distribution patterns, suggesting a cell-to-cell propagation [7, 8]. Recently, it is also found that some of the neurodegeneration associated proteins can be detected in peripheral organs and may also present concomitantly in the brain and peripheral tissues [9]. These identified molecular pathological backgrounds of the disease-associated proteins along with the inconsistent clinical symptoms of NDDs create a necessity of proper neuropathological examinations like developments of biomarkers, clinical and neuroimaging studies which finally lead to the accurate diagnosis [9]. The treatment of these neurodegenerative diseases are mostly symptomatic such as dopaminergic treatment for PD and movement disorders, anti-inflammatory and analgesic for neuronal infections and pain, cholinesterase for cognitive disorders, antipsychotic for dementia, etc. though, further progress in therapeutic management is needed to treat many other progressive and serious symptoms of the diseases [10–12]. Integrative treatments along with medicinal therapies are also in the frontline of research to improve the endogenous antioxidant systems targeting the oxidoreductase enzymes and thereby the activity of daily life of the neurodegenerative patients. These integrated treatments act by protecting against oxidative and nitrosative stress related neuropsychiatric disorders, sensory and other symptoms of non-motor fluctuations, fatigue, etc. [11]. In this chapter, the

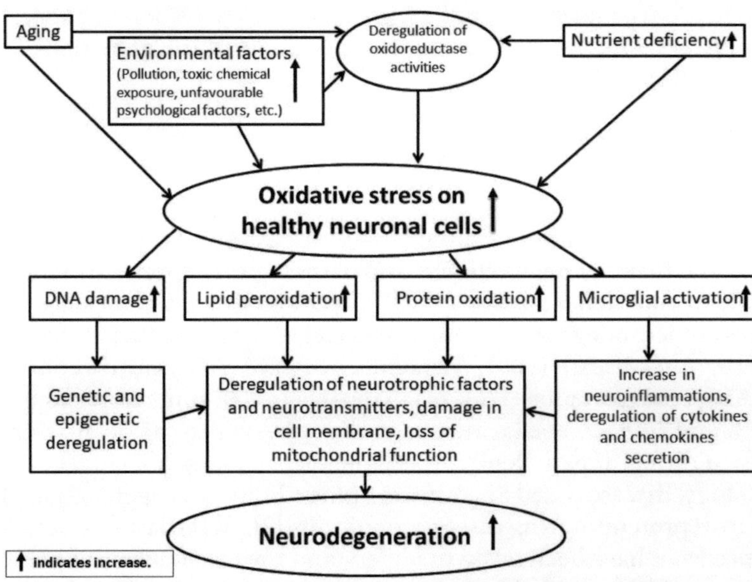

Figure 1.
Schematic presentation of possible steps for the action of different factors involve in the development of neurodegeneration.

diagnostic classification of NDDs, their preventive strategy, and treatment with a special emphasis on the oxidoreductase enzymes are summarized to understand the current progress in the field of NDDs.

2. Role of oxidative stress in neurodegenerations

While aging is the key contributor to most of the NDDs, oxidative stress is the main factor for functional impairment during aging due to the oxidation of lipids, deoxyribonucleic acid (DNA), and proteins in presence of reactive oxygen or nitrogen species (ROS or RNS). Thus, it is not unreasonable to assume that enhancement in level(s) of ROS and/or RNS increase(s) the senescence of cells by secreting pro-inflammatory factors and enzymes followed by cellular degradation [13] (**Figure 1**). S-Nitrosylation reaction plays a crucial role in nitric oxide (NO) bioactivity and is shown to have neuroprotective as well as a neurotoxic role based on the targeted protein [14]. An increase in level of nitrosative stress may affect mitochondrial respiration by inhibiting its complexes I and IV and disrupts the mitochondrial dynamics followed by synaptic injury and neuronal damage [15]. Thus, it may be corroborated that this RNS mediated protein modification is associated with AD pathology, as AD can be characterized by increasing mitochondrial dysfunction [16, 17]. On the other hand, an increase in the level of amyloid-beta (Aβ) and aging aggravate the senescent phenotype and endothelial cell dysfunction and can be characterized by oxidative stress [13]. It is well proved that reduction in oxidative stress can reduce the cognitive impairment and inflammatory processes as oxidative stress enhances the loss of homeostasis [6]. Increased level of oxidative stress also enhances the production of the inflammatory cytokine and finally both affect the cognitive performance in aged individuals [4]. In this context, it is obvious to mention that the involvement of oxidoreductases in oxidative stress is a well-accepted logic-based fact in NDDs [18, 19].

2.1 Role of oxidoreductase in NDDs

Oxidoreductases are the enzymes that catalyze the oxidation–reduction reactions by transferring electrons from oxidant to reductant. It can be classified as oxidases, dehydrogenases, peroxidases, hydroxylases, oxygenases, and reductases. It has been found that increased levels of oxidative stress biomarker glutathione peroxidase (GSH-Px) and reduction in its (GSH-Px) activity are associated with an increase in inflammatory cytokines and both of them has a correlation with the cognitive impairment of elderly individuals [4]. Increased expression of nuclear factor erythroid2-related factor 2 (Nrf2) and reduced level of superoxide dismutase 1 mRNA are associated with cognitive impairments [20]. Nrf2 is the main controller of oxidative response and toxic insults to cells and modulate the expression of the inflammatory, metabolism-related gene [21]. Signaling pathways such as glycogen synthase kinase 3 (GSK-3), nuclear factor kappa light chain enhancer of activated B cells (NF-κB), NOTCH, and adenosine monophosphate kinase (AMP kinase) and Kelch ECH associating protein 1 (Keap1) regulates the Nrf2 activity [6, 22]. It has been observed that Nrf2 deficiency along with amyloidopathy and tauopathy induce neuroinflammation and oxidative stress providing a direct connection between neurodegeneration and oxidoreductase system [23]. Harada et al. [18] have shown a positive association of the NQO2 (dihydronicotonamide riboside (NRH): quinone oxidoreductase 2, or QR2) and PD as the deletion of 29-bp nucleotides in the promoter region of the NQO2 gene associates with the development of PD. In presence of catechol quinones, the over-expression of

NQO2 in brain cells leads to the production of ROS (via the rapid conversion of superoxide radicals into hydrogen peroxide and then into highly reactive hydroxyl radicals)-induced neuronal cell death or neurodegeneration [5]. The other isoform of NAD(P)H:quinone acceptor oxidoreductase (NQO), the NQO1, or NADH quinone oxidoreductase of mitochondria carries the most common Leber's hereditary optic neuropathy (LHON) mutants [24]. The protein disulfide isomerase (PDI) enzyme is another potent oxidoreductase resides in the endoplasmic reticulum, has the ability to catalyze the oxidative folding reactions requires for the maturation of disulfide-bond-containing proteins. It is found to regulate the molecular trafficking along the secretory pathway to prevent the protein misfolding which can mitigate the proteinopathy-induced neurodegenerative diseases (e.g., AD, PD). Monoamine oxidase (MAO) is another oxidoreductase which is predominantly found in the brain regional and platelet mitochondrial outer membrane catalyzes the amine (-NH_2) compound (monoamine neurotransmitters e.g., serotonin (5-HT), dopamine) and formaldehyde and hydrogen peroxide (H_2O_2) as byproducts. During aging the MAO-A activity has been found to be increased in cerebral cortex, hippocampus, hypothalamus, and pons-medulla [25, 26] whereas, decrease in blood platelets [27]. Very limited information are available there about the aldo-keto oxidoreductase (aldehyde dehydrogenase or ALDH, aldose reductase, aldehyde reductase, alcohol dehydrogenase) which can detoxify the reactive aldehyde and ketone bodies in the brain bearing a protective role from the development of aging-induced neurodegenerative diseases, especially AD. The oxidative damage to the polyunsaturated fatty acids (PUFA) generates the 4-hydroxy-trans-2- nonenal (HNE) and its related carbonyl, which creates immunoreactivity. Their elevated levels are found in the brain as well as in cerebrospinal fluid (CSF) of AD, PD, and ALS patients [28]. It (reactive aldehydes) inhibits mitochondrial functions, disrupts cytoskeleton, inhibits glutamate transporters, and also modifies tubulin structure [29, 30]. The aldo-keto oxidoreductases have the ability to detoxify these reactive aldehydes in brain by converting those into corresponding acid or alcohol [31, 32]. The ALDH has been found in the cerebral cortex, hippocampus, basal ganglia, and midbrain, and aldose reductase in the pyramidal cells of cerebral cortex and hippocampal CA1 region, while all of these oxidoreductases are present in cerebellum [19] providing and strengthening the evidence of the fact that cerebellum is less vulnerable in the proteinopathy related NDDs. The xanthine oxidoreductase which converts hypoxanthine to xanthine and thereby to uric acid-producing H_2O_2 as a byproduct can generate superoxide via NADH oxidase activity and similar to ALDH, manganese superoxide dismutase (Mn-SOD) and heme-oxygenase-1 (HO-1) are promptly expressed in reactive astrocytes and found to be present in healthy pyramidal neurons [33, 34].

3. Types of neurodegeneration

NDDs often overlap with each other based on pathology and symptoms especially in multisystem atrophy where several areas get affected at a time making it difficult to analyze clinically [35]. Based on the predominant pathological features and topography of the central nervous system (CNS) during the diseased condition, NDDs have been classified into three major aspects:

3.1 Anatomical classification

The anatomical positions (such as cerebral cortex, basal ganglia, brainstem, cerebellum, spinal cord) in relation to the disease condition (**Table 1**) can be used as

DOI: http://dx.doi.org/10.5772/intechopen.94950

Disease	Main anatomic vulnerability	Symptoms	Main neuropathology	Protein aggregate(s)	Diagnostic approaches	Therapeutic strategies	References
Alzheimer's disease	Basal forebrain, Frontal and Temporal lobes, Limbic structures, Locus coeruleus and Olfactory bulb	Cognitive and functional impairment, Dementia like memory loss, Problems with abstract thinking, Planning, Flexibility, Motor tasks, Neuropsychiatric manifestations and Language problem	Neurofibrillary tangles (NFTs), Neuropil threads, Neuritic and amyloid plaques and Amyloid angiopathy	Aβ 3R+ 4R tau	Anatomical distribution of (a) neuronal tau pathology, (b) extracellular Aβ deposits and (c) CAA	iAβ5 (Chaperon) for inhibiting protein aggregates, Donepezil and Rivastigmine drug therapy, APP regulation by latrepirdine and treatment with cholinesterase inhibitors and HDACi	Finkel, 2004 [36]; Desai and Grossberg; 2005 [10]; Okun et al., 2004 [37]
Parkinson's disease	Substantia nigra pars compacta, Trans-entorhinal region, Motor and Sensory cortex, Prefrontal cortex, Dorsal motor nuclei of the medulla oblongata, Raphe nucleus and Locus coeruleus of the brainstem	Motor symptoms: Tremor (resting), Muscle rigidity, Postural instability, Coordination problem, Slow movements, Bradykinesia and Loss of physical movement, Non motor symptoms: High-level cognitive dysfunction, Psychiatric and emotional changes, Depression, Difficulty in swallowing and speaking, Sensory symptoms and Constipation and/or Urinary problems	Neuronal degeneration of dopaminergic neurons	α-synuclein	Estimation of the activity of terminal dopa decarboxylase (DDC), Evaluation of the availability of presynaptic dopamine transporters (DAT) and Vesicular monoamine transporter 2 (VMAT2) density measurements in dopamine terminals.	Combination of Levodopa and Carbidopa, Inducers of Hsp104 chaperones, Targeting of α- synuclein misfolding with Hsp 70, Treatments with anti-inflammatory drugs against Methyl-4-phenylpyridinium induced autophagy and Knockdown of Sirt2 by siRNA	Brooks, 2005 [38]; Djaldetti et al., 2006 [39]; Quinn, 1995 [40].

Disease	Main anatomic vulnerability	Symptoms	Main neuropathology	Protein aggregate(s)	Diagnostic approaches	Therapeutic strategies	References
Amyotrophic lateral sclerosis	Motor cortex, Brainstem motor neurons and Spinal cord motor neurons	Progressive muscle atrophy, Fasciculation (muscle twitching), Spasticity and Hyporeflexia	Upper and lower motor neuron loss, Bunina bodies, Neuronal inclusions and Astrocytic hyaline inclusions	TDP-43	Morphology and subcellular distribution of protein deposits in neurons and Anatomical distribution of protein deposits	Vitamin E therapy to reduce oxidative stress	Strong, 2003 [41]
Huntington's disease	GABAergic medium spiny neurons (MSNs) in the striatum	Dystonia (involuntary limb movement), Incoordination, Cognitive decline and Behavioral disturbances	GABAergic neurons	Huntingtin	Determining cerebral blood flow (both its decrease and increase) and local brain metabolism, Change in dopamine receptor expression,	Dopamine receptor blockers (e.g. phenothiazines), Targeting of mHTT misfolding with Hsp70, Immunomodulation therapy and Rapamycin-induced autophagy	Andrews and Brooks, 1998 [42].
Prion's Disease	Pyramidal in HP, and Granular neurons in DG	Dementia, Difficulties in walking and speaking, Fatigue, Muscle stiffness, Hallucination and Confusion	Spongiform changes and Prion protein (PrP) accumulation	PrP	Morphology of PrP deposition, Glycosylation pattern and electrophoretic mobility of PK-resistant PrP by using western blot technique, Codon 129 polymorphism and Aetiology if known	RNAi-mediated silencing of host-encoded cellular prion protein (PrPC)	Kovacs and Budka, 2009 [8].

DOI: http://dx.doi.org/10.5772/intechopen.94950

Disease	Main anatomic vulnerability	Symptoms	Main neuropathology	Protein aggregate(s)	Diagnostic approaches	Therapeutic strategies	References
Multiple sclerosis (MS).	Superior medial frontal cortex, Superior dorsolateral frontal cortex, Medial occipital lobe, Lateral occipital cortex, Deep inferior parietal white matter, and Pons	Depression, Fatigue, Anxiety, Personality change, Tremor, Unilateral loss of vision, Pain, Bladder problems, Constipation and Impaired hearing	Inflammation and Demyelination	TDP-43, SOD1, FUS and DPRs	Morphological, subcellular and anatomical distribution of protein deposits	Immunomodulation by beta-interferon, Ocrelizumab etc and Hormonal replacement therapy	Berger and Reindl, 2007 [43]

3R+ 4R tau: 3 or 4C-terminal microtubule binding repeats in Tau protein; APP: amyloid precursor protein; Aβ: amyloid-beta; CAA: Cerebral amyloid angiopathy; DG: Dentate gyrus ; DPRs: dipeptide repeat proteins; FUS: Fused in sarcoma gene; HDACi: histone deacetylase inhibitors; HP: Hypothalamus; Hsp: Heat shock protein; iAβ5: 5-residue β sheet breaker peptide; mHTT: mutant huntingtin; siRNA: Small interfering RNA; Sirt2: sirtuins 2; SOD1: Superoxide dismutase 1; TDP-43: Transactive response (TAR) DNA-binding protein 43; VMAT2: vesicular monoamine transporter 2.

Table 1.
Different types of neurodegenerative diseases and their respective information.

a component to identify the disease and also for its classification [7]. For example, dementia is a pathological condition due to neurodegeneration in the cerebral cortex as observed in AD patients. Similarly abnormal motor functions as observed in PD are associated with degenerations involving basal ganglia including nucleus putamen, globus pallidus, substantia nigra, subthalamic nucleus, red nucleus, and some thalamic and brainstem nuclei, etc. (**Table 1**) [7].

3.2 Based on conformational and biochemical modifications of proteins

Some proteins and their cellular aggregation as identified, are associated with NDDs and found to undergo conformational and biochemical modifications during disease pathology [7]. Proteins such as microtubule-associated protein Tau encoded by MAPT on 17q21 chromosome, Aβ transcript encoded by AβPP gene on chromosome 21q21.3, α- Synuclein encoded by a gene (SNCA) on chromosome 4, prion protein (PrP), encoded by a gene (PRNP) on chromosome 20, Transactive response (TAR) DNA-binding protein 43 (TDP-43) encoded by the TARDBP gene on chromosome 1, etc. are few of the examples. Some hereditary associated proteins encoded by genes that are associated with neurological trinucleotide repeat disorders like ataxins, huntingtin, atrophin-1 are also found as a biomarker of disease identification [44]. Protein deposition pattern in CNS during NDDs are classified into several proteinopathies such as cerebral amyloidoses, tauopathies, α-synucleinopathies, prion diseases, trinucleotide repeat diseases, TDP-43 proteinopathies, FUS/FET proteinopathies, neuroserpinopathy, etc. [7, 44–47]. Only a few numbers of modifications are so far included in the classification and pathological subtyping of the NDDs, e.g. Aβ modification is not included in the classification of AD but the biochemical steps of Aβ aggregation and a different variant of Aβ aggregates have implemented to interpret the early and late phases of AD pathology [48]. Similarly, recent neuropathological studies have revealed that the biochemical classification of tauopathies by analyzing the insoluble and trypsin-resistant tau with varying C-terminal fragments [49]. Tauopathies can be further distinguished by the presence of different ratios of the repeat (R)- and 4R-tau and two or three major phospho-tau bands (60, 64, and 68 kDa) [50]. Other major protein modifications are shown in **Table 2**.

3.3 Cellular pathology

Neurodegenerative diseases can be characterized by the presence of misfolded proteins within or outside of the neurons [51]. Cellular pathology is also an important aspect to distinguish the location of protein deposition at the subcellular level such as nuclear, neuritic (axons or dendrites), cytoplasmic, mitochondria, myelin, lysosomes or in astrocytes, etc. [7]. Charcot–Marie–Tooth neuropathy type 1B (CMT-1B), a hereditary motor and sensory neuropathy, is one such example as it can be identified by the accumulation of misfolded myelin protein zero (mpz) in the endoplasmic reticulum [51]. In superoxide dismutase 1 (SOD1) associated ALS, misfolded SOD1 mutant is found in the cytosol [51]. An aggregate of huntingtin with an expandable polyQ track in the cytosolic and nuclear space of the neuronal cells is a characteristic of HD [52]. α-synuclein, a soluble protein marker of PD is available predominantly in the presynaptic zone of neuronal cells. The complex mechanism of AD consists of binding of Aβ oligomers with several receptors intracellularly. Accumulation of Aβ in the lysosomal compartment followed by a change in its membrane permeability is also identified in AD pathological. Aβ-induced mitochondrial dysfunction by the deregulation of enzymatic activities of the electron transport chain is also identified in AD pathology [52].

Proteinopathies	Proteins	Biochemical characteristics	References
Amyloidoses	Amyloid-beta (Aβ)	• Produced by proteases mediated the sequential cleavage • Most abundant component of Aβ deposit is Aβ 1–40/1–42 peptides along with other available species such as (Aβ 1–37/38/39) • Aβ deposits have resistance against proteinase K	Thal et al., 2015 [48]
	Prion protein (PrP)	• Disease-associated PrPSc is detergent-insoluble and resistance to protease K treatment but not the physiological cellular form of PrP (PrPC) • PrP can be differentiated based on electrophoretic mobility and N-terminal sequence of the core fragments, and the most common PrPres species is PrP27–30. • Other fragments forms are PrP 11, PrP7–8, PrP14, PrP-CTF12/13, PrP16–17, and PrP17.5–18	Kovacs and Budka, 2009 [8]; Duyckaerts et al., 2009 [45]
Tauopathies	Tau	• Most common modification is hyperphosphorylation • Ratio between 3R- and 4R-tau, and two or three major phospho-tau bands (60, 64, and 68 kDa) in Western blot of sarkosyl-insoluble fractions are the major factors for distinguishing tauopathies. • Distinct feature of taupathies are N- and C-terminal truncation, glycation, nitration of tyrosine residues, glycosylation, transglutamination, deamidation; acetylation; oligomer; the banding patterns of C-terminal fragments of tau and the trypsin-resistant band patterns etc.	Lee et al., 2001 [50]; Taniguchi-Watanabe et al., 2015 [49]
Synucleinopathies	α-Synuclein	• Modification occur: phosphorylation at serine 87 and 129 and at tyrosine 125 residue • Various conformation and oligomeric states of synuclein are in dynamic equilibrium state. • Resistance against protease K	Dehay et al., 2015 [46]
TDP-43 Proteinopathies	Transactive response (TAR) DNA-binding protein 43 (TDP-43)	• Modification: phosphorylation on serine 379 (S379), S403, S404, S409, S410 residues • Ubiquitinylation and abnormal cleavage; oligomer; C-terminal fragments detected in disease	Kovacs, 2019 [47]

Aβ 1–40/1–42/1–37/38/39 are different amyloids oligomers consisting different numbers of residue-long proteolytic fragments; PrP27–30,PrP 11, PrP7–8, PrP14, PrP-CTF12/13, PrP16–17, and PrP17.5–18 are different fragments of prion protein consisting of different size (Kda) of fragments; Prp-CFT: C-terminal fragments of PrP; PrPSc: abnormal or scrapie isoform of PrP; 3R- and 4R- Tau: 3 or 4C-terminal microtubule binding repeats in Tau protein.

Table 2.
Biochemical characteristics of some major proteins related to neurodegenerative proteinopathies.

4. Diagnosis of NDDs

The NDDs can be largely differentiated by the anatomical regions showing neuronal dysfunction, biochemical and conformational changes in protein markers and neuronal cell pathologies including the deposition of protein(s), and alteration in genetics and epigenetics [53]. Structural neuroimaging techniques such as computed tomography (CT), magnetic resonance imaging (MRI) are used for diagnosis but due to very low specificity, they have been replaced by new neuroimaging techniques such as positron emission tomography (PET) and single-photon emission computed tomography (SPECT) [54]. The functional magnetic resonance imaging (fMRI), another new generation diagnostic approach, identify the correlation of different physiological functions during NDDs rather than direct imagining of neuronal activities [55]. The focus of the new diagnostic researches is to establish easily detectable biomarkers from blood or saliva to distinguish the different forms of neuronal disorders [56]. It has been observed by using fluorodeoxyglucose-PET that the earliest sign of AD is metabolic decline. Detection and identification of small molecule metabolites in the biological samples are called metabolomics, which is another newest cutting edge approach for the diagnosis of neurodegenaration associated metabolic disorders [57]. Genetic markers associated with familiar neurodegenerative diseases are already identified for a different type of disorders such as for diagnosis of AD amyloid precursor protein, presenilin gene mutations and apolipoprotein E (APOE) polymorphism are some of the known genetic markers whereas for PD, α-synuclein protein or PrP gene mutation for the familiar type of prion diseases, etc. have been identified, although their sensitivity and specificity are still questionable [58].

4.1 Application of neuro-imaging in the diagnosis of major diseases due to neurodegenerations

The most frequent CNS diseases are diagnosed by using the following functional neuroimaging techniques:

4.1.1 Parkinson's disease (PD)

PD is known to be progressive as well as a degenerative disorder associated with a loss of dopamine-producing neurons of the substantia nigra and other brain regions [38]. Pathophysiology, progression, and complications of this disease are well understood and identified by neuroimaging techniques. Neuroimaging deals with the detection of the changes in brain structure as well as its region(s) on the basis of changes in brain glucose, oxygen and dopamine metabolism, and receptor binding of dopamine [38]. The functional markers such as (a) the activity of dopa decarboxylase (DDC) terminal, (b) presynaptic dopamine transporters (DAT) availability, and (c) vesicle monoamine transporter density in dopamine terminals (VMAT2) are implemented for neuroimaging (in both PET/SPECT) [38]. DDC works as a catalyst for L-Dopa decarboxylation to Dopamine. Using 6-[18F]-L-dopa PET the activity of DDC can be measured by measuring neuronal loss. 18F-Dopa transfer into 18F-dopamine by amino acid decarboxylase and trapped in synaptic vesicles, whose uptake depends on the presence or loss of nigrostriatal postsynaptic dopamine cell [39, 59]. Similarly, DAT, which helps to clears dopamine after its release in the synaptic cleft, can be used for PD diagnosis. D2–dopamine receptor binding tracers 11C-raclopride-PET and 123I-iodobenzamide (IBZM)-SPECT have been used for the assessment of D2 receptor density and gives good results to evaluate PD patients [38, 39, 59]. VMAT2 is an integral membrane protein which especially transports dopamine like monoamines into synaptic vesicles. 11C-dihydro-tetrabenazine-PET can be used for its test [38]. In PD loss of 5-HT concentration is observed by 11C-WAY100635-PET and the measurement of 5-HT1A receptor by evaluating the functional integrity of serotonergic neurons [59]. fMRI analysis of PD patients has shown the distinct variation in covariance patterns of the region-based resting-state activity in functional brain regional networks in comparison to the normal brain. The detrimental effect of dopamine replacement on non-motor brain functions due to the alteration of the physiological pattern of dopamine signaling can also be proved by fMRI studies [60]. This suggests functional changes between the three different brain-related disorders. PD can be characterized by the activation of the neuroimmune system in microglia followed by a loss of neurons in substantia nigra [61]. 11C-PK11195 is known to enable the detection of increased signals in substantia nigra which reflect local degeneration as a consequence of PD [38]. In addition, the significant reduction in metabolomes like catecholamines [homovanillic acid (HVA), dihydroxyphenylacetic acid (DOPAC), L-dopa, etc.) has also been observed during PD [62]. NMR metabolomics based study has helped to differentiate PD from non-PD patients by detecting the presence of metabolomes (like creatinine, glucose, lactate, 3-hydroxyisobutyric acid and 3-hydroxyisovaleric acid etc.) in CSF [62]. The presence of kynurenine in the blood of PD is proved to be potential biomarker candidates [62].

4.1.2 Alzheimer's disease (AD)

Structural neuroimaging with CT and volumetric MRI has an application on AD related cerebral atrophy and measurement of cerebral blood flow or regional glucose and oxygen metabolism [63]. MRI helps to measure the memory forming zone of CNS i.e. hippocampus and cortex-structures in the temporal lobe and further helps to differentiate between AD and other dementia [64, 65]. By using magnetic resonance spectroscopy (MRS) the information about concentrations of tissue substrate or metabolite during AD and MCI (mild cognitive impairment) can be identified by using N-acetyl aspartate as a marker [64, 66]. Quantification of amyloid deposition by tracing the amounts of radioligands *in vivo* is also possible by PET and SPECT. PET usually detect the metabolic uptake of fluorine 18 [18F]-labeled 2 fluorodeoxyglucose (2-deoxy-2-[18F]- fluoro-D-glucose- FDG) and blood flow in patients with dementia [64, 67]. The fMRI techniques in AD diagnosis is implemented in cerebral blood flow (CBF) and cerebral vasomotor regulation (CVR) mapping. This limitation in CVR has been observed in the APOEε4 gene carrying early-onset AD patients with vascular dysfunction, which occur due to the astrocytic end-feet swelling, degeneration of pericyte, hypertrophy of basement-membrane as well as due to the abnormalities in the endothelial-cell metabolic. Non-invasive fMRI is a major tool for the diagnosis of AD by identifying such changes in CVR mapping [55]. In the AD brain, the most characteristic feature and useful biomarker are amyloid plaques consisting of Aβ protein, dystrophic neuritis, inflammatory factors, and cellular material inside and outside of the neurons [68]. Tau tangles are also associated with AD and composed of paired helical filaments (PHF) derived from abnormally hyperphosphorylated microtubule-associated protein tau [69]. Radiotracers such as [18F]-BAY94–9172, an Aβ ligand, have been used with PET to differentiate between AD and frontotemporal dementia patients [66, 70]. PET studies with the application of [11C] PIB, a derivative of thioflavin-T amyloid dye that binds to Aβ plaques but not tangles, show more retention in the cortical zone of frontotemporal dementic brain when compared to AD brain [64, 65]. 18F-DDNP – PET scanning helps to compare AD, MCI, and controls having intact cognitive functions [71, 72]. The plasma metabolomics biomarkers including glycerophosphatidylcholines, asparagine, acylcarnitines, and asymmetric dimethylarginine (ADMA) are identified as a predictive marker of plasma which can predict the risk of conversion from cognitively normal individuals to AD [57]. Reduction in N-acetyl aspartate in the brain can be correlated with neuronal and mitochondrial dysfunction during AD. Acylcarnitine, sphingomyelins, glycerophospholipids found to be increased significantly in the CSF of AD patients in comparison to normal patients [57].

4.1.3 Huntington's disease (HD)

HD is a dominantly inherited, autosomal, NDD characterized by motor, cognitive, and emotional abnormalities [42]. In the early course of HD, no structural changes of the brain can be observed by CT and MRI while only in later stage atrophy has been observed in the caudate and frontal cortex [73]. PET study can provide information by diagnosing HD as early as 9 to 11 years before the first symptoms appear [74]. PET with 2-deoxy-2-[fluorine-18]fluoro-D-glucose (^{18}F-FDG-PET) is also used to detect the reduced striatal glucose metabolism in early HD which further causes bradykinesia, dementia, and putamen hypometabolism connects with chorea and eye-movement abnormalities [42, 75]. HD has also been found to be associated with structural loss of dopamine (D1, D2) receptor-expressing medium spiny neurons from the striatum. The damage can be estimated by using radiolabelled dopamine antagonists [11C] raclopride and by observing the binding potential (BP) of dopamine receptors which help to assess the neuronal damage [42]. Further, PET study using [11C] diprenorphine as a tracer has shown a mild loss of opioid receptors in the striatum in HD patients [42]. The accumulation of active microglia due to neuronal loss can be seen with the help of 11C-(R)-PK11195 as a tracer in the striatum, globus pallidus, and frontal cortex in HD patients [76]. fMRI has applied to diagnosed HD by various cognitive paradigm including maze learning, serial reaction time, working memory etc. fMRI blood-oxygen-level-dependent (BOLD) signal response is also applied for correlation between different regions in HD patients [74]. HD is associated with metabolic and energy pathways alterations. After studying various metabolomics Mastrokolias et al. [77] have found that the deregulation of phosphatidylcholine metabolism is a prominent plasma biomarker of HD.

4.1.4 Amyotrophic lateral sclerosis (ALS)

ALS is a motor neuron disease (MND) associated with progressive deterioration of the corticospinal tract, brainstem, and anterior horn cells of the spinal cord [78]. Cortical atrophy is observed in late ALS which can be assessed by structural MRI of ALS patients' CT studies. The increased population of microglia during ALS can be observed by radiolabelled PET ligand [11C] (R)- PK11195 which selectively binds with the peripheral benzodiazepine binding site (PBBS) of microglia [79]. ALS can be also diagnosed by the measurement of postsynaptic dopamine D2 receptor binding abilities. 123I-benzamide (123I-IBZM), a specific binding substance with D2 receptors shows less receptor binding during ALS when investigated using SPECT [80, 81]. PET studies show a decrease in 11C-flumazenil (a radiolabelled antagonist of benzodiazepine receptor) binding in the primary sensory, premotor, prefrontal, thalamic, and parietal regions during ALS [78, 82]. Both 123I-N-isopropyl-p-iodoamphetamine (123I-IMP) and 99mTc-hexamethyl propylene amine oxime ([99mTc] –D, L- HMPAO) are markers which have been used to determine reduced fronto-temporal blood flow as well as glucose metabolism by SPECT studies [41].

4.1.5 Multiple sclerosis (MS)

MS is characterized by demyelination of neurons in the CNS, with the formation of plaques or lesions [43]. MRI studies show these lesions are dynamic in different stages of the disease. Neuronal loss and brain atrophy are not visible in the early stage of MS by MRI scans [83]. Decreased regional and global CBF and cerebral metabolic rate of glucose (CMRglc) can be observed by PET imaging using 18FDG as a detection agent. Although the differentiation between acute and chronic MS is tough for the SPECT study with the help of Tc-99 m-MIBI as a radiopharmaceutical, which in fact shows multiple accumulation points in acute MS but not in chronic MS [84]. Binding potential of microglial peripheral benzodiazepine binding sites (PBBS) towards [11C] (R)-PK11195 can be applied as a determinant factor of MS, like other neurodegenerative diseases [83, 85]. Application of fMRI in MS has been recently applied to assess MS-associated modification of cervical cord in the patient. This study also helps to identify the brain regions involved in the tactile and proprioceptive stimulation during AD pathology [86]. Mangalam et al. [87] have performed a study to find out the MS-based untargeted metabolic alterations in bile acid biosynthesis as well as the metabolism of histidine, taurine, tryptophan linoleic acid, and d-arginine.

4.2 Anatomical identifications of neuronal losses in relation to clinical symptoms

Identification of anatomical positions is needed for understanding the early symptoms. For example, brain regions such as the entorhinal cortex, neocortex, hippocampus, limbic system are responsible for symptoms like cognitive decline, dementia, and other high-order brain functions alterations whereas basal ganglia, thalamus, brain stem, and motor cortical areas are mostly responsible for disturbance in body movements. Combinations of these types of symptoms are mostly observed during the progression of diseases following region-specific neurodegenerations [7]. One of the conventional approaches to understand neuroanatomy is brain mapping which helps to localize the disease-related changes [88, 89]. Characterization of brain regional anatomical changes in diseased conditions is a fascinating and advanced approach of current neuroscience research [90]. Earlier the identification of neurofibrillary tangles in the cortex and the hippocampus due to Alzheimer's were studied by region-of-interest technique (ROI). These techniques (ROI) can be implemented to compute the overall volume for a particular brain structure, based on the manual or automatic positioning of the sections' MRI serially for a subject by using already existing anatomical protocols [91]. Though the prior knowledge of anatomical structures makes ROI-based analysis a strong and useful approach, the lack of detailing in the investigation of the underlying complex structure makes this method less advantageous for diagnosis [92]. Like in AD patient ROI technique is capable of establishing the hippocampus and entorhinal cortex as most prominent imaging biomarkers but this technique is not useful to investigate the underlying complex structure of the hippocampus for further diagnosis [91–93]. Another newer image analysis technique is Voxel-based morphometry (VBM) able to identify cortical and subcortical degeneration simultaneously, providing significant insight changes in gray matter in AD and MCI [94]. VBM can be implemented to classify MRI maps into individual maps of gray matter, white matter, and CSF tissue classes followed by creating an alignment of gray matter maps and then smoothened it with the help of filters. The corresponding cognitive score has been statistically assessed using multiple regression analysis. The general linear model used to fit with gray matter density at each image location or voxel related to diagnosis, cognitive scores, etc. [95, 96]. Several brain regional gray matter atrophy such as temporal, posterior cingulated, precorneal cortex in AD and normal aged persons has been documented by using VBM studies [97]. The application of VBM has also been observed to investigate the effect of aging and gender on spatial profiling in normal subjects [96] as well as in the frontotemporal zone of PD, and Lewy body dementia, and also in herpes simplex encephalitis [98–100].

The limitation of VBM is inherently low spatial resolution due to spatial smoothening to achieve inter-individual cortical variability [101]. VBM study is also not optimal for analysis of gray matter atrophy as highly convoluted features that appeared for the Gyral and sulcal region cannot be readily distinguished leading to a lack of detecting and localizing the subtle cortical differences [89].

4.3 Brain mapping: a diagnostic tool for neurodegenerative diseases

Brain mapping techniques rely on a mathematical computation of anatomy where brain surface and its volumes are represented as 3D complex geometrical patterns mesh models, averaged, combined across subjects and can be statistically defined [89]. The technique implies transformable and deformable templates which can be transformed into brain shape for studies by constraining surface landmarks (e.g., sulci) or alignment of surface-specific geometrical patterns (e.g., gyri) and helps to co-localized cortical and subcortical regions along with cortical thickness, gray matter density, functional activations, etc. by improving the identification of cortical and subcortical changes associated to the diseases [102]. Localization of changes by cross-sectional and longitudinal imaging is done by tensor-based morphometry (TBM), a newer approach to map the changes in the brain over time. This method has been found to be sensitive, with high throughput, and attractive for gauging brain changes in larger study populations. In cross-sectional studies, where many individual images are matched to a common brain template to compare the systematic volume and shape between control and diseased individuals, TMB has been found to be effective and helps to clinically correlate the different disease conditions like Fragile X syndrome 56 and Williams syndrome, etc. [103]. TBM can detect and visualize subcortical nuclear as well as structural gray and white matter by using newer statistical methods [103].

4.3.1 Application of brain mapping in AD

Radial atrophy mapping of the hippocampus has been first applied by Thompson et al. [104] for the diagnosis of AD and has shown distinct differences between normal elderly and AD patients. Later on, Frisoni et al. [105] have demonstrated that the CA1 area and parts of the subiculum using the same technique and showed AD cases have 15–20% atrophy in relation to the normal controls. Apostolova et al. [88] have shown that patients with MCI have more severe involvement of CA1 and subiculum atrophy which are likely to convert into AD at a later age. Apolipoprotein E4 (APOE ε4) a prominent genetic risk factor for sporadic AD carriers, has a higher hippocampal atrophy rate than non-carriers as observed by longitudinal MRI study. Further, Bookheimer et al. [106] by cortical thickness study have shown that cognitively normal APOE ε4 carriers have a significantly thinner entorhinal cortex and focal hippocampal atrophy in comparison to normal non-carriers. Thompson et al. [104] have also reported based on the comparison of baseline grey matter density map a significant atrophy in lateral, temporal, parietal, and parieto-occipital cortices in AD patients. Brain mapping by computational anatomy techniques has significantly improved sensitivity for the detection of differences in the disease-induced groups. The study between amnestic MCI and mild AD subjects has shown a highly significant greater cortical atrophy in AD patient despite a small cognitive difference between these two groups [88]. It has also been observed that in sporadic early-onset of AD (EOAD; <65 years of age) and late-onset of AD (LOAD; >65 years of age), subjects can also be differentiated by severity and localization of cortical atrophy. While EOAD shows widespread atrophic changes, LOAD subjects show lower rate and more focal pattern of entorhinal, para-hippocampal, inferior temporal, posterior cingulate/precuneal, and lateral temporal changes, suggesting younger AD subjects have displayed higher cognitive reserve and tolerance to pathological burden as cortical neurodegeneration correlates with cognitive declines [105].

4.3.2 Application of brain mapping in dementia

Dementia with Lewy bodies (DLB) is associated with some of the features such as cognitive decline, early-onset hallucinations and delusions, Parkinsonism, and a fluctuating course. Pathologically hallmarks for DLB are synuclein-rich intracellular deposits known as Lewy bodies is often observed as a hallmark of DLB in patients as well as in few cases of AD. It has been observed that a distinct cortical atrophy pattern i.e. hippocampal and inferior temporal preservation along with midbrain atrophy occurs in DLB but not in AD when studied by VBM [100]. Ballmeier et al. [107] have mentioned that the preservation of the temporal and orbitofrontal cortices in demented subjects is also a distinct feature of DLB.

5. Preventive measures for neurodegenerative diseases

The estimated number of total dementia cases globally is around 50 million among which 60% of the cases are from low or middle-income countries and also 10 million new cases are reported globally every year as per the recent report of WHO. Hence, the demands of health care and social services are huge and need constant surveillance to decrease the rate of incidence of this type of life-threatening diseases as well as its associated expenses. It has been observed that up to 10 years before the diagnosis of dementia, cognitive impairment is likely to appear in individuals and it declines sharply in the final stage of 3 years [108]. Individuals with deficits in vitamin B12, folate, and thyroid-stimulating hormones (TSH) are found to involve with poorer cognitive performances [109]. Elevated levels of serum-homocysteine and cardiovascular diseases are also responsible for cognitive impairments [110]. Depressed mood, hip fracture, polypharmacy, history of psychoses are the reasons behind cognitive impairment without dementia (CIND) in older age and the low education, depression, APOE ε4 allele, medicated hypertension, midlife elevated serum cholesterol, and high diastolic pressure, as well as diabetes and anticholinergic medication, are responsible factors for mild cognitive impairment (MCI) [111, 112]. The strongest risk factor of dementia and AD is age and lifetime cumulative multiple risk factors like genetic susceptibility, environmental exposure, and biological factors etc. are also needed to be considered for identification of preventive measures [113, 114]. For example, genetic and environmental factors are responsible for Familial AD as reported by many and found to be happening in 58% of AD cases [114, 115]. The involvement of APOE ε4 allele as a genetic factor as well as some other genes in AD is well established and, APOE polymorphism can partially explain the familial aggregation of AD in 15–20% of AD cases which generally affect 75 years or older patients [116, 117]. Vascular risk factors and AD or dementia are also found to be associated and the control of amendable vascular disease-associated risk factors has found to offer preventive measures for AD [118]. Such as controlling high blood pressure, diabetics, and mid-life obesity are important interventions and found to show a better score in cognitive tests and reduce the risk of dementia and AD in very old individuals [118]. One more important aspect is psychosocial factors and it has been reported that attending higher education in early-life, work complexity in adult life, intellectually stimulating activities are also found to help in delaying the onset of dementia [119]. Physical activities have a beneficial effect on mental health [120]. It has been reported by Alzheimer's associations [120] that mentally, physically and socially active lives have the potential to postpone the onset of clinical dementia by 5 years and substantially decrease the number of dementia cases in the community.

5.1 Role of diets and micronutrients in the prevention of neurodegenerative diseases

Dietary components are found to be effective in the prevention of neurodegenerative diseases [121]. It has been observed that docosahexaenoic acid (DHA), an n-3 polyunsaturated fatty acid, enriched diet such as fish (fatty or blue species), shellfish, and algae [122] plays a relevant role in the preservation of histopathology of the neuronal tissue and helps in memory and learning maintenance [123]. Apart from that, polyphenols, curcumin like food components have neuroprotective properties [124, 125]. Polyphenols are a natural antioxidant and show activities on chelation, scavenging free radicals, survival gene activations, cell signaling pathways, and also regulating mitochondrial function by the ubiquitin-proteasome system [124]. On the other hand, curcumin is an anti-amyloid drug and responsible for reducing oxidation of protein and also reduce pro-inflammatory cytokines interleukin-1beta in AD-induced transgenic mice brains [125]. Deficiency of vitamin B, C, and E are associated with AD development [126]. Some mixed results in this regard have been observed in several clinical studies [127, 128]. As reported, a lower level of dietary and supplemented folic acid is associated with AD pathology and lack of folic acid due to malabsorption and malnutrition can increase the chance of AD by two folds in the elderly [127, 129]. Some studies have found no significant role of Vitamin B9, B12, C, E in AD pathology, whereas the other study gives evidence that Vitamin C and E have a neuroprotective effect on AD [128]. It has also been studied that while sufficient intake of vitamin E, omega-3 fatty acid and omega-6 fatty acid, vitamins A, C, and whole grains increase neuronal activation, food component like saturated fatty acids, cholesterol and sodium significantly lower the neuronal activation and gray matter volume [130]. One more significant effect of diet on neuronal health is observed when a calorie-restricted diet is consumed by aged individuals [131]. It has observed that calorie restriction augments brain-derived neurotrophic factor (BDNF), induces sirtuins a silent information regulator proteins responsible for the regulation of life span, repair, and protection of DNA, etc. [132]. While some dietary components' actually have the preventive function on neurodegeneration on the other hand some obesity-induced dietary components act oppositely and increase the risk of AD, PD, and neuroinflammatory disease [133].

5.2 Physical exercise for prevention of neurodegenerative diseases

Physical exercise for short term or long term has been found to be beneficial on neurodegenerations and cerebrovascular diseases as observed in both animal and human model [134]. Physical exercise and the expression of different neurotrophic factors [like BDNF, insulin like growth factor-1 (IGF-1), vascular endothelial growth factor (VEGF)] are found to be associated, and hence promote neural plasticity and neurogenesis in the hippocampus [135]. It has also been observed that upregulation of BDNF in circulation as well as in the brain can be induced by exercise which can be corroborated with an increase in cognitive function [135]. Due to exercise metabolite-like ketone bodies accumulates in the hippocampal region of the brain which alters BDNF promoter and promotes BDNF expression [136]. BDNF expression is regulated by various genetic factors and pathways like Val66Met mutation [137], PGC-1α/FNDC5 pathway [138], APOEε4 allele carriers [139] and methyl CpG binding protein 2, etc. [140]. Reports are also available on the effect of exercise-induced increase level of VEGF in relation to the reduction in ischemic injury and improvement in cognitive performance. This may be due to an increase in progenitor cell proliferation and all the cell differentiation in ischemic penumbra [141]. Some data has depicted that exercise-induced muscular VEGF increases the level of VEGF in the hippocampus. However, this VEGF helps in neurogenesis and angiogenesis in the ischemic brain followed by improved cognitive activity [142]. The muscle-derived IGF-1 has been found to increase IGF-1 permeability via BBB by increasing the IGF-1 receptor expression in BBB followed by IGF-1 concentration in the hippocampus possibly by regulating IGF binding proteins (IGF-BPs) [143]. Alteration of cytokine production by exercise can also restore the IGF-1 level followed by a reduction in neurodegeneration [143]. Physical exercise, as reported, when used as an adjuvant therapy of psychotropic drugs gives better anti-depressant and anti-anxiety outcome and also effectively reduces dementia [144]. Production of several myokines in muscles (like PGC-1α, Irisin and Cathepsin B, etc.) is promoted by exercise. Myokines that are beneficial for the brain (fibroblast growth factor 21 (FGF-21) and SPARC etc.) also generate after exercise. Serotonin (5-HT) concentration generally increases after exercise in serum, whole blood, and also in urine [145]. This peripheral 5HT level is found to be correlated with an increased level of 5-HT in the brain [146]. Exercise is also associated with a decrease in AD specific deposition of Aβ and tau pathology in the brain [147]. High-intensity exercise provides an improvement in PD associated impaired motor functions [148]. Exercise can also reduce the loss of dopaminergic neurons and fibers and decrease α-synuclein in the nigrostriatal region as observed in animals [149].

6. Therapeutic strategies

Conventional therapies such as cholinesterase inhibitors for AD or Levo-dopa for PD provide symptomatic relief but not on effective disease progression. Advancement in the knowledge of the neuro-molecular mechanism of NDDs has helped to develop new drugs to counteract pathological aggregation of the protein [150]. The prevention of abnormal protein aggregation or targeting of misfolded proteins for their degradation by new therapeutic agents is a potential field of recent researches [151]. Reports are also available on the inhibition of fibril formation by several compounds like antibodies, molecular chaperones, nanoparticles of polyphenols, metal chelators, and tetracyclines nanoparticles by inhibiting the aggregating pathways of different amyloidogenic proteins, such as Aβ, α-synuclein, PrP protein, etc. [152]. But the most challenging part of therapeutic strategies are (a) prevention of the formation of oligomers or converting aggregation process into alternative non-toxic pathways (b) transporting the therapeutic agents through the blood–brain-barrier (BBB) and (c) delivery of nanoparticulate drugs to the targeted neuron to reduce dose-dependent toxicities and side effects.

6.1 Available treatment paradigm

Among the few food and drug administration (FDA) approved drug regimen Donepezil and Rivastigmine like acetylcholine esterase inhibitors are used as palliative treatment which help to reduce the progression of AD but not for the long-term [36, 153]. A combination of levodopa and carbidopa has been successfully delivered via the BBB to treat PD patients. The drugs act by converting into dopamine after decarboxylation in the substantia nigra of the PD patient and increase the level of dopamine in that zone for the first few years of treatment after consecutive consumption [40]. Dopamine agonists such as Pergolide, Bromocriptine Parlodel also have therapeutic efficacy but show cardiovascular and endocrinological problems [154]. In HD, reserpine, or dopamine receptor blockers (i.e. phenothiazines) are used to impair dopamine transport and to reduce overactivity in dopaminergic nigrostriatal pathways [155]. In MS, the preventive measures taken to combat the relapse are prednisone to reduce inflammation, beta-interferon, Ocrelizumab, glatiramer acetate, alemtuzumab, mitoxantrone for immunomodulation, and Ocrelizumab for reducing the primary progression [156].

6.2 Future strategies

Effective treatment of NDDs needs identification and mitigation of risks, complete cure, or at least a long term relief from the symptoms that need to be achieved despite all the advancement of genetic, biomolecular, and pharmaceutical sciences. Some of the strategies are designed or may execute in lower animals but clinical trials and human applications are yet to be achieved. Some novel therapeutic strategies are summarized below.

6.2.1 Inhibition of disease-associated protein deposition

Protein misfolding are occurred due to gene mutations, oxidative stress, aging, altered cellular temperature, pH, etc. [157]. The misfolded proteins are often partially unfolded by molecular chaperones and then go through a self-rearrangement to form oligomeric aggregates that are finally converted into amyloid fibrils [158]. Accumulations of such protein aggregates are often found to be in relation to amyloidosis of the CNS as well as symptoms of neurodegeneration [159]. Synthetic chaperons or short peptides with a recognition motif of misfolded proteins have therapeutic potential to disaggregate this protein aggregation during amyloidosis. Heat shock protein 104 (Hsp104) one of the major molecular chaperones, has been found to be efficacious to disaggregate proteins aggregates in yeast cells [160]. Hsp104 has shown to eliminate various amyloid conformations and reduce deposits of pre-amyloid oligomers by binding with the protein fibril and blocking the aggregation process [160]. One variant of Hsp104 has properties to dissolve α- synuclein aggregates in the PD model. Hsp70 and its related compound also have neuroprotective roles as found by in-vitro and in-vivo studies (**Figure 2**) [160]. Hsp70 is also responsible for restoring Tau homeostatic [161]. In-vitro and in-vivo studies have revealed that Hsp70 and related compounds help to clear Aβ depositions and restore Tau homeostasis [160, 161]. The proposed therapeutic strategies are tabulated in **Figure 2**. HD is associated with polyQ induced misfolding of mHTT mutant Huntington which can be the target and inhibit by Hsp70 [162]. Curcumin or epigallocatechin gallate have Hsp70 simulating properties resulting in a reduced level of neuronal death followed by improved cognitive and motor deficits [163]. On the other hand, α-synuclein misfolded protein aggregation is associated with PD, and

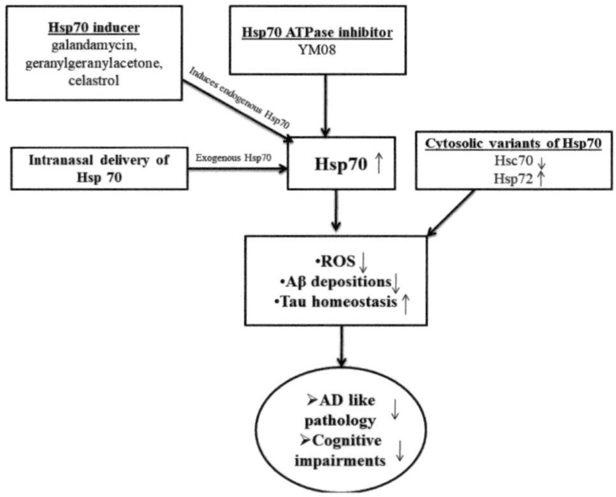

Figure 2.
Different strategies for inducing Hsp 70 as molecular chaperone in AD brain.
Molecular chaperones of heat shock protein 70 have neuroprotective properties such as, maintenance of the tau homeostasis, and decrease in the reactive oxygen species (ROS) generation and amyloid-beta (Aβ) depositions which are related to both AD pathology and cognitive impairments. There are other cytosolic isoforms of Hsp 70 i.e. Hsc 70 and Hsp 72 whose downregulation and upregulation respectively also show similar effect on the above mentioned parameters. Neuroprotective effect of Hsp 70 can be achieved by (a) targeting Hsp 70 ATPase using its inhibitor, YM08, (b) Hsp 70 inducers (for example, galandamycin, geranylgeranylacetone, and celastrol) or (c) intranasal delivery of Hsp 70. ↓ and ↑ indicate increase and decrease respectively.

Hsp70 can bind with α-synuclein and halt the further misfolding of α- synuclein followed by refolding [164]. In this context, it may be mentioned that specific Hsp and related chaperones significantly contribute to targeting protein aggregates (mHTT, α-synuclein, etc.) specific to NDDs, and the development of these Hsp stimulant can be beneficial for future therapies [165].

6.2.2 Neuroimmunomodulatory therapies

Neuronal death and neurodegeneration are in connection with the vicious cycle of inflammation especially when inflammatory mediators stay in the tissue for a longer period. AD, PD, and HD cases express higher plasma and CSF concentrations of proinflammatory cytokines, such as IL-6, TNF-α, IL-1β, IL-2, IL-6, and cyclooxygenase-1/2, etc. [166] whereas, anti-inflammatory cytokines and growth factors (IL-10, TGF-β, CD206, etc.) producing microglia becomes lower in number in such patients [167]. The monocytes isolated from the carriers of the HD gene, express the mutant Huntingtin protein and show hyperactivity to lipopolysaccharide stimulations [168]. Thus, it may be concluded that the hyperactive immune system is an important feature of HD pathogenesis and its associated immunomodulators can be used for potential HD treatment. The presence of CD8+ and CD4+ peripheral lymphocytes in substantia nigra has been found in post mortem brains of PD patients [169]. Anti-inflammatory drugs such as minocycline, resveratrol, tanshinone, and silymarin have therapeutic promises against PD by blocking the activation of NADPH oxidase and microglial activation and pro-inflammatory cytokine release [12, 157]. Apart from that, it has also been investigated that the monoclonal antibodies against the α-synuclein not only reduced protein propagation and amyloid formation [170] but also ameliorated dopaminergic neuronal cell loss and improved PD-like pathologies, followed by improving motor deficits in PD induced mouse [170].

6.2.3 Autophagy

Neuronal cells are subjected to autophagy but they have a very strong lysosomal system which is effective for the removal of protein aggregates and dysfunctional mitochondria as well as the rapid removal of the autophagosomes [171]. The autophagy regulating gene mutation often leads to NDDs like AD, ALS, and PD as a consequence of large gathering of the debris of dysfunctional organelles and undiluted waste proteins [172]. The factors responsible for the suppression of autophagy followed by suppressing neuronal functions and plasticity are stress signals, hypoxia, mechanical damages, decreased level of amino acids, etc. [173]. Mutation of autophagy-related genes has shown neurodegeneration in lower animals like mice and flies [174, 175]. A promising therapeutic strategy is a drug-induced autophagy in neurogenerative patients for the removal of abnormal proteins as observed in animal models. One of the recent examples of such drug-induced autophagy is methyl-4-phenylpyridinium (MPP+) which induces apoptosis in dopaminergic neurons by disrupting the complex I of the electron transport chain of mitochondria of mouse Parkinson's Models [176]. Similarly, the antihistaminic drug, Latrepirdine shows a regulation of APP in the AD mice model [177]. Drugs like calcium channel blockers and USFDA approved rapamycin show potential to stimulate the autophagic process followed by the clearance of mutant huntingtin protein in lower animals [178]. Rapamycin is also able to reduce Aβ-induced cognitive deficits of AD by activation of the AMPK-mTOR signaling pathway in aged as well as Type 2 Diabetes Mellitus-induced AD cases [179]. mTOR signaling has the ability to form autophagic vacuoles, mitigating tau and Aβ deposition and controlling the apoptotic pathways. Metformin has been found to involve in autophagy by AMPK-dependent mechanism of HD as well as dephosphorylates neurofibrillary tangles of tau in AD and is established as a potential therapeutic agent for NDDs [177].

6.2.4 Neurotrophic factors and possible strategies for neurogenesis

Dysregulation of the neurotrophic factors which are the molecular aids of neuronal functions such as differentiation, growth, etc., is associated with NDDs [180]. The affected region of the brain starts losing neurons and glia in absence of functional regulation of these molecules [181]. Some Factors such as nerve growth factors (NGF), BDNF have the ability to bind with tyrosine receptor kinases, inhibit apoptotic signals, and promotes cell survival by promoting tissue growth by cell proliferation [182] and also their absence play a prominent role in neurodegenerative diseases [183]. A decreased level of NGF in AD patients induces cellular death followed by loss of neuronal functions whereas a decrease of BDNF in substantia nigra does the similar in PD patients due to degeneration of synaptic connections [184]. An increase in the level of these neurotrophic factors in the degenerated brain regions could be a possible therapeutic strategy although their larger size and polar nature make them unsuitable for transport through BBB and difficult to target [185] although gene delivery injection and neurotrophin mimetics are already under investigations [183]. Another important marker of NDDs is the deficiency of neurosteroids during AD [37], PD [37], HD [186], and MS [187] which can be defended by hormonal replacement therapy and found to be beneficial in AD, PD, HD, and MS patients [188].

6.2.5 Insulin associated neurodegeneration

Insulin signaling in CNS is responsible for differentiation, proliferation, neurite growth, and shows neuroprotective as well as anti-apoptotic activity [189]. The structural and functional integrity of synapses, neurons, and neuronal circuits followed by memory and learning are depend partially on Insulin and affected by diabetes mellitus and metabolic syndromes [189]. Insulin resistance in the brain is also associated with an increased level of phospho-Tau and Aβ42 [189]. Evidence is there of decrease in levels of insulin in the brain and CSF during AD with an increase in Aβ42 and advanced glycation [189]. Insulin administration enhances Aβ42 clearance and improves working memory and cognition [189]. Insulin receptor-associated genes IRS-1 pSer616 and IRS-1 pSer636/639 have been identified in relation to Aβ oligomer levels and function as a biomarker for AD [190]. Antidiabetic drug Metformin is reported to inhibit cognitive decline which may have some connection with the insulin signaling pathway in CNS although needs further investigation [191].

6.2.6 Cholinergic system in AD

The connection between the cholinergic system and AD has been hypothesized as presynaptic cholinergic markers are found to be depleted in the cerebral cortex during AD pathology [192], nucleus basalis of Meynert (nbM) in the basal forebrain undergoes severe degeneration in AD [193], and memory gets weaken by the cholinergic antagonist while agonists have the opposite effect [194]. Cholinesterase inhibitors such as donepezil, rivastigmine, and galantamine have been found to improve significantly the cognitive activity related to AD [195].

6.2.7 Targeting oxidoreductases

A very limited information are available about targeting oxidoreductases to inhibit the proteinopathies and consequent neurodegeneration. Among the free radical generator, the NOX bears a significant role in oxidative stress-induced neurodegeneration. The pharmacological NOX inhibitors have been found to improve different NDDs and it is well-reviewed by Barua et al. [196]. The PDI, as discussed before is associated with different proteinopathies (like AD and PD) and its attenuation could be a promising approach to counteract the proteinopathy-induced neurodegeneration. Polyphenols curcumin, from a turmeric (*Curcuma longa*) spice and masoprocol (from *Larrea tridentata*), have found to restore the ROS-induced chaperone damage, protein misfolding, and thereby neurodegenerative disease, sustaining traffic along the ER's secretory pathway by preserving functional integrity of PDI [197].Nrf-2 also an important transcription factor, associated with the oxidoreductase system is also a crucial target to deal with. Naringin (4′,5,7-trihydroxy flavonone 7-rhamnoglucoside), the flavonone found in grapefruit and related citrus species has been found to upregulate the Nrf-2 and its consequent cytoprotective genes to act as a neuroprotective molecule [198]. Carnosine, an endogenous dipeptide biomolecule has been recently found to be a potent inhibitor of aging-induced increase in brain regional monoamine oxidase-A activity [26] and it can also reduce and restore the aging-induced deposition of amyloid-beta plaque quantitatively as well as qualitatively [199, 200]. Interestingly, the inhalation of patchouli oil (extracted from the leaf of *Pogostemon cablin*) has also the ability to modulate the blood platelet MAO-A activity and thereby in mood behavior [201].

6.2.8 Epigenetic modulations

Epigenetic markers such as histone deacetylases have been proved to be involved with AD. Treatment with HDACi (histone deacetylase inhibitors) such as sodium butyrate, phenylbutyrate, suberoylanilide hydroxamic acid, resveratrol induces phosphorylation of tau protein, reduces the amyloidogenic processing of APP followed by restoration of learning and memory deficits in AD patients [202]. PD is also associated with epigenetic modulation. Where sporadic PD patients are linked with α-synuclein hypomethylation in dopaminergic neurons, the familial PD patients show a decrease in histone acetylation followed by an increase in α-synuclein levels [203]. α-synuclein-mediated neurotoxicity has been found to reduce by the treatment with Sirt2 siRNA [204]. A decrease in PD symptoms with the administration of dopamine may also be correlated with the deacetylation of histone H4 lysine 5 (H4K5), histone H4 lysine 12 (H4K12), and histone H4 lysine 16 (H4K16) [205].

7. Conclusion

The NDDs are progressive neuronal cell deaths due to environmental, biochemical, genetic and epigenetic factors. Generation of free radicals due to the oxidoreductase activity and deterioration of antioxidant system are found to trigger the aggregation of misfolded proteins in CNS causes mitochondrial dysfunctions and neuro-inflammations which finally leads to NDDs (**Figure 3**) [2]. Based on the predominant pathological features, NDDs can be classified in three different way, i.e. (a) anatomical, (b) the proteins undergoing conformational and biochemical modifications and (c) cellular pathology. Protein deposition pattern in CNS during NDDs are classified into several proteinopathies such as (a) cerebral amyloidoses, (b) tauopathies, (c) α-synucleinopathies, (d) prion diseases, (e) trinucleotide repeat diseases, (f) TDP-43 proteinopathies, (g) FUS/FET proteinopathies, (h) neuroserpinopathy, etc. depending upon the major protein aggregates [8, 44–46]. The diagnosis of neurodegenerative diseases is mostly associated with quantification of their specific receptor binding, changes in cellular metabolism or in anatomical structure. The neuroimaging techniques such as PET, SPECT, fMRI

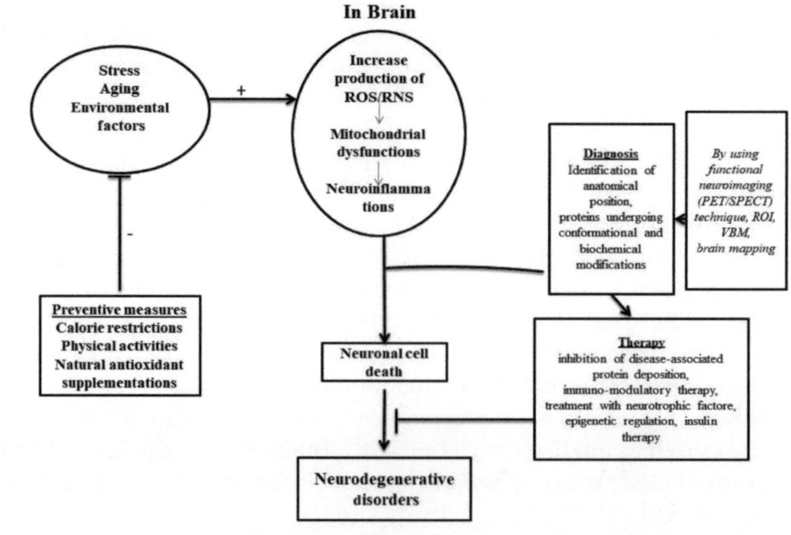

Figure 3.
Schematic representation of overall causes, diagnosis, prevention and therapy of NDDs.
The preventive measures are able to prevent the early causes of NDDs, while the diagnosis and therapy at later stage target and try to control the diseased condition due to NDDs. ROS: Reactive oxygen species; RNS: Reactive nitrogen species; PET: Positron emission tomography; SPECT: Single photon emission computed tomography; ROI: Region of interest; VBM: Voxel-based morphometry. + and - indicate activation and inhibition respectively.

etc. have been extensively used to diagnosed receptor activities and metabolic faith of damaged neuronal cells during diseased condition by using radio labeled tracers. Applications of metabolomics is another newer approach for the diagnosis and prognosis of NDDs. On the other hand, the characterization of brain regional anatomical changes in diseased conditions can be performed by brain mapping techniques. These advancements of technologies made the diagnosis of neurodegeneration much easier and an early diagnosis is also possible to some extent for most of the major NDDs. Although complete cure from NDDs/neurodegenerative disease(s) is not yet achieved but therapies that can prevent the early occurrence of NDDs are investigated. Individuals' deficits of vitamin B12, folate, and thyroid-stimulating hormones (TSH), cardio vascular and metabolic disorders, genetic and environmental factors are few of the reason behind NDDs and can be prevented by taking proper measure from the early life. Physical exercise, calorie restriction and few dietary components like DHA, polyphenols have neuroprotective effect and found to be beneficial for NDDs. Apart from prevention, there are limited medicated therapies are available in the market for the treatment of NDDs. But many strategies like inhibition of disease-associated protein deposition, immunomodulatory therapy, treatment with neurotrophic factor, epigenetic regulation, targeting oxidoreductases, and insulin therapy are under investigations and clinical trials. The current advancement in biochemical, pathological and pharmaceutical researches may ensure a better future of global neuronal health but it needs adaptation of a healthier lifestyle from the early day of life to avoid the occurrence of such NDDs.

Acknowledgements

The authors are thankful to the Department of Pharmaceutical Technology, Jadavpur University, Kolkata; University Grants Commission (UGC), New Delhi; Science and Engineering Research Board (SERB), Govt. of India; Indian Council of Medical Research (ICMR), New Delhi, India; Department of Psychology, Neuroscience Program, Central Michigan University, Mount Pleasant, Michigan, USA for their support to prepare the present work.

Conflict of interest

There is no conflict of interest to declare.

Author details

Mrinal K. Poddar[1]*, Apala Chakraborty[1] and Soumyabrata Banerjee[2]

1 Department of Pharmaceutical Technology, Jadavpur University, 188, Raja S. C. Mallick Road, Kolkata, 700032, India

2 Department of Psychology, Neuroscience Program, Central Michigan University, Mount Pleasant, MI, 48859, USA

*Address for all correspondences to: mrinalkp@yahoo.com

References

[1] Przedborski S, Vila M, Jackson-Lewis V. Neurodegeneration: What is it and where are we? J. Clin. Invest. (2003) 111:3-10.

[2] Jellinker KA. Basic mechanisms of neurodegeneration: acritical update. J Cell Mol Med. (2010) 14:457-487.

[3] Hengartner MO. The biochemistry of apoptosis. Nature (2000) 407:770-776.

[4] Melo A, Monteiro L, Lima RMF, de Oliveira DM, de Cerqueira MD et al. Oxidative stress in neurodegenerative diseases: mechanism and therapeutic perspectives. Oxid Med Cell Longev. (2011) 2011:467180.

[5] Cassagnes L-E, Chhour M, Pério P, Sudor J, Gayon R, Ferry G, Boutin JA, Nepveu F, Reybier K. Oxidative stress and neurodegeneration: The possible contribution of quinone reductase 2. Free Rad Biol Med (2018) 120: 56-61.

[6] Morroni F, Sita G, Graziosi A, Ravegnini G, Molteni R, et al. PQM130, a novel feruloyl-donepezil hybrid compound, effectively ameliorates the cognitive impairments and pathology in a mouse model of Alzheimer's disease. Front Pharmacol. (2019) 10:658.

[7] Kovacs G. Molecular pathological classification of neurodegenerative diseases: turning towards precision medicine. Int J Mol Sci. (2016) 17:189.

[8] Kovacs GG, Budka H. Molecular pathology of human prion diseases. Int J Mol Sci. (2009) 10:976-999.

[9] Dugger BN, Hoffman BR, Scroggins A, Serrano GE, Adler CH et *Al. tau* immunoreactivity in peripheral tissues of human aging and select tauopathies. Neurosci Lett. (2019) 696:132-139.

[10] Desai AK, Grossberg GT. Diagnosis and treatment of Alzheimer's disease. Neurology (2005) 64:S34–S39.

[11] Chaudhuri KR, Schapira AHV. Non-motor symptoms of Parkinson's disease: dopaminergic pathophysiology and treatment. Lancet Neurol. (2009) 8:464-474.

[12] Rees K, Stowe R, Patel S, Ives N, Breen K et al. Non-steroidal anti-inflammatory drugs as disease-modifying agents for Parkinson's disease: Evidence from observational studies. Cochrane Database Syst Rev. (2011): Cd008454.

[13] Liguori I, Russo G, Curcio F, Bulli G, Aran L, et al. Oxidative stress, aging, and diseases. Clin Interv Aging. (2018) 13:757-772.

[14] Nakamura T and Lipton SA. S-Nitrosylation of critical protein thiols mediates protein misfolding and mitochondrial dysfunction in neurodegenerative diseases. Antioxid Redox Sign. (2011) 14:1479-1492.

[15] Cho D-H, Nakamura T, Fang J, Cieplak P, Godzik A, et al. S-Nitrosylation of Drp1 mediates beta-amyloid-related mitochondrial fission and neuronal injury. Science. (2009) 324:102-105.

[16] Abtahi SL, Masoudi R, Haddadi M. The distinctive role of Tau and amyloid beta in mitochondrial dysfunction through alteration in Mfn2 and Drp1 mRNA levels: a comparative study in *Drosophila melanogaster*. Gene (2020) 754:144854.

[17] Wang W, Zhao F, Ma X, Perry G, Zhu X. Mitochondria dysfunction in the pathogenesis of Alzheimer's disease: recent advances. Mol Neurodegener. (2020) 15:30.

[18] Harada S, Fujii C, Hayashi A, Ohkoshi N. An Association between Idiopathic Parkinson's Disease and Polymorphisms of Phase II Detoxification Enzymes: Glutathione *S*-Transferase M1 and Quinone Oxidoreductase 1 and 2. Biochem Biophys Res Commun. (2001) 288: 887-892.

[19] Picklo Sr MJ, Olson SJ, Markesbery WR, Montine TJ. Expression and Activities of Aldo-Keto Oxidoreductases in Alzheimer Disease. J Neuropathol Exp Neurol (2001) 60: 686-695.

[20] Mota SI, Costa RO, Ferreira IL, Santana I, Caldeira GL et al. Oxidative stress involving changes in Nrf2 and ER stress in early stages of Alzheimer's disease. Biochim Biophys Acta. (2015) 1852:1428-1441.

[21] Butterfield DA, Keller JN. Antioxidants and antioxidant treatment in disease. Biochim Biophys Acta Mol Basis Dis. (2012) 1822:615.

[22] Deshmukh P, Unni S, Krishnappa G, Padmanabhan B. The Keap 1–Nrf2 pathway: promising therapeutic target to counteract ROS-mediated damage in cancers and neurodegenerative diseases. Biophys Rev. (2017) 9:41-56.

[23] Rojo AI, Pajares M, Rada P, Nuñez A, Nevado-Holgado AJ, et al. NRF2 deficiency replicates transcriptomic changes in Alzheimer's patients and worsens APP and TAU pathology. Redox Biology. (2017) 13:444-451.

[24] Carelli V, Rugolo M, Sgarbi G, Ghelli A, Zanna C, Baracca A, Lenaz G, Napoli E, Martinuzzi A, Solaini G. Bioenergetics shapes cellular death pathways in Leber's hereditary optic neuropathy: a model of mitochondrial neurodegeneration. Biochim Biophys Acta. (2004) 1658: 172-179.

[25] Adolfson R, Gottfries CG, Oreland L, Wiberg A, Winblad B. Increasedactivity of brain and platelet monoamine oxidase in dementia of Alzheimer type. Life Sci. (1980) 27: 1029-1034.

[26] Banerjee S, Poddar MK. Carnosine: effect on aging-induced increase in brain regional monoamine oxidase-A activity. Neurosci Res (2015) 92: 62-70.

[27] Banerjee S, Poddar MK Platelet monoamine oxidase-A activity and aging: effect of carnosine. J Physiol Sci (2013) 63:279-285.

[28] Witz G. Biological interactions of α,β-unsaturated aldehydes. Free Rad Biol Med (1989) 7: 333-349.

[29] Keller JN, Mark RJ, Bruce AJ, et al. 4-Hydroxynonenal, an aldehydic product of membrane lipid peroxidation, impairs glutamate transport and mitochondrial function in synaptosomes. Neurosci (1997) 80: 685-696.

[30] Neely MD, Sidell KR, Graham DG, Montine TJ. The lipid peroxidation product 4-hydroxynonenal inhibits neurite outgrowth, disrupts neuronal microtubules, and modifies cellular tubulin. J Neurochem (1999) 72: 2323-2333.

[31] Mitchell D, Petersen D. The oxidation of α,β unsaturated aldehydic products in lipid peroxidation by rat liver aldehyde dehydrogenases. Toxicol Appl Pharmacol (1987) 87:403-410.

[32] O'Connor T, Ireland LS, Harrison DJ, Hayes JD. Major differences exist in the function and tissue-specific expression of human aflatoxin B1 aldehyde reductase and the principal human aldo-keto reductase AKR1 family members. Biochem J (1999) 343: 487-504.

[33] Furuta A, Price DL, Pardo CA, et al. Localization of superoxide dismutases in Alzheimer's disease and Down's syndrome neocortex and hippocampus. Am J Pathol (1995) 146: 357-367.

[34] Schipper HM, Cisse S, Stopa EG. Expression of heme oxygenase-1 in the senescent and Alzheimer-diseased brain. Ann Neurol (1995) 37: 758-768.

[35] Burn DJ, Jaros E. Multiple system atrophy: cellular and molecular pathology. Mol. Pathol. (2001) 54:419-426.

[36] Finkel SI. Effects of rivastigmine on behavioral and psychological symptoms of dementia in Alzheimer's disease. Clin Ther (2004) 26: 980-990.

[37] Okun MS, DeLong MR, Hanfelt J, Gearing M, Levey A. Plasma testosterone levels in Alzheimer and Parkinson diseases. Neurology. (2004) 62: 411-413.

[38] Brooks D. Imaging studies in drug development: Parkinson's disease. Drug Discov Today Technol. (2005a) 2:317-321.

[39] Djaldetti R, Ziv I, Melamed E. The mystery of motor asymmetry in Parkinson's disease. Lancet Neurol. (2006) 5:796-802.

[40] Quinn N. Drug treatment of Parkinson's disease. BMJ Br. Med J. (1995) 310: 575-579.

[41] Strong MJ. The basic aspects of therapeutics in amyotrophic lateral sclerosis. Pharmacol Therapeut. (2003) 98: 379-414.

[42] Andrews TC, Brooks DJ. Advances in the understanding of early Huntington's disease using the functional imaging techniques of PET and SPET. Mol Med Today. (1998), 532-539.

[43] Berger T, Reindl M. Multiple sclerosis: disease biomarkers as indicated by pathophysiology. J Neurol Sci. (2007) 259: 21-26.

[44] Ou SH, Wu F, Harrich D, Garcia-Martinez LF, Gaynor RB. Cloning and characterization of a novel cellular protein, TDP-43, that binds to human immunodeficiency virus type 1 TAR DNA sequence motifs. J. Virol. (1995) 69:3584-3596.

[45] Duyckaerts C, Delatour B, Potier MC. Classification and basic pathology of Alzheimer disease. Acta Neuropathol. (2009) 118:5-36.

[46] Dehay B, Bourdenx M, Gorry P, Przedborski S, Vila M, et al. Targeting _-synuclein for treatment of Parkinson's disease: Mechanistic and therapeutic considerations. Lancet Neurol. (2015) 14:855-866.

[47] Kovacs GG. Molecular pathology of neurodegenerative diseases: principles and practice. J Clin Pathol. (2019) 72:725-735.

[48] Thal DR, Walter J, Saido TC, Fandrich M. Neuropathology and biochemistry of Aβ and its aggregates in Alzheimer's disease. Acta Neuropathol. (2015) 129:167-182.

[49] Taniguchi-Watanabe S, Arai T, Kametani F, Nonaka T, Masuda-Suzukake M, et al. Biochemical classification of tauopathies by immunoblot, protein sequence and mass spectrometric analyses of sarkosyl-insoluble and trypsin-resistant tau. Acta Neuropathol. (2016) 131:267-280.

[50] Lee VM, Goedert M, Trojanowski JQ. Neurodegenerative tauopathies. Annu. Rev. Neurosci. (2001) 24:1121-1159.

[51] Bertolotti A. Importance of the subcellular location of protein deposits

in neurodegenerative diseases. Curr Opin Neurobiol. (2018) 51:127-133.

[52] Ruz C, Alcantud JL, Vives Montero F, Duran R, Bandres-Ciga S. Proteotoxicity and Neurodegenerative Diseases. Int J Mol Sci. (2020) 21:5646.

[53] KovacsGG.Introduction:Classification of neurodegenerativediseases. In Neuropathology of Neurodegenerative Diseases: A Practical Guide. Kovacs GG, Ed; Cambridge University Press: Cambridge UK. (2015), 1-8.

[54] Camprodon JA, Stern TA. Selecting neuroimaging techniques: a review for the clinician. Prim Care Companion CNS Disord. (2013) 15:PCC.12f01490.

[55] Chen JJ. Functional MRI of brain physiology in aging and neurodegenerative diseases. Neuroimage. (2019) 187:209-225.

[56] Malamud D. Saliva as a diagnostic fluid. Dent Clin North Am. (2011) 55:159-178.

[57] Wilkins JM, Trushina E. Application of Metabolomics in Alzheimer's Disease. Front Neurol. (2018) 8:719.

[58] García JC, Bustos RH. The Genetic Diagnosis of Neurodegenerative Diseases and Therapeutic Perspectives. Brain Sci. (2018) 8:222.

[59] Seibyl J, Chen W, Silverman D. Single-photon emission computed tomography and positron emission tomography evaluations of patients with central motor disorders. Nucl Med. (2008) 03:274-286.

[60] Siebner HR, Meder D, Herz DM. fMRI in Parkinson's Disease. (2020) In: Ulmer S, Jansen O. (eds) fMRI. Springer Nature Switzerland AG.

[61] Wang S, Jing H, Yang H, Liu Z, Guo H, et al. Tanshinone I selectively suppresses pro-inflammatory genes expression in activated microglia and prevents nigrostriatal dopaminergic neurodegeneration in a mouse model of Parkinson's disease. J Ethnopharmacol. (2015) 164: 247-255.

[62] Shao Y, Le W. Recent advances and perspectives of metabolomics-based investigations in Parkinson's disease. Mol Neurodegeneration. (2019) 14:3.

[63] Honson NS, Johnson RL, Huang W, Inglese J, Austin CP, et al. Differentiating Alzheimer disease-associated aggregates with small molecules. Neurobiol Dis. (2007) 28: 251-260.

[64] Small GW, Bookheimer S, Thompson PM. Current and future uses of neuroimaging for cognitively impaired patients. Lancet Neurol. (2008) 7:161-172.

[65] Craig-Schapiro R, Fagan AM, Holtzman DM. Biomarkers of Alzheimer'sdisease. Neurobiol Dis. (2009) 35:128-140.

[66] Petrella J, Coleman E, Doraiswamy P. Neuroimaging and early diagnosis of Alzheimer disease: a look to the future. Radiology. (2003) 226:315-336.

[67] Jagust W. Molecular neuroimaging in Alzheimer's disease. NeuroRx. (2004) 1: 206-212.

[68] Schipper HM. The role of biologic markers in the diagnosis of Alzheimer's disease. Alzheimers Dement. (2007) 3:325-332.

[69] Hampel H, Goernitz A, Buerger K. Advances in the development of biomarkers for Alzheimer's disease: from CSF total tau and A 1-42 proteins to phosphorylated tau protein. Brain Res Bull. (2003) 61:243-253.

[70] Waragai M, Okamura N, Furukawa N, Tashiro M, Furumoto S

et al. Comparison study of amyloid PET and voxel-based morphometry analysis in mild cognitive impairment and Alzheimer's disease. J Neurol Sci. (2009) 285:100-108.

[71] Cai L. Synthesis and evaluation of two 18F-Labeled 6-Iodo-2-(4c-N,N dimethylamino) phenylimidazo [1,2-a] pyridine derivatives as prospective radioligands for a-amyloid in Alzheimer's disease. J Med Chem. (2004) 4: 2208-2218.

[72] Kulkarni PV, Arora V, Roney AC, White C, Bennett M et al. Radiolabeled probes for imaging Alzheimer's plaques. Nucl Instrum B. (2005) 241: 676-680.

[73] Kloppel S, Henley SM, Hobbs NZ, Wolf RC, Kassubek J, et al. Magnetic resonance imaging of Huntington's disease: preparing for clinical trials. Neuroscience. (2009) 164: 205-219.

[74] Paulsen JS. Functional imaging in Huntington's disease. Exp Neurol. (2009) 216: 272-277.

[75] Wong FC, Kim EE. A review of molecular imaging studies reaching the clinical stage. Eur J Radiol. (2009) 70: 205-211.

[76] Tai YF, Pavese N, Gerhard A, Tabrizi SJ, Barker RA, et al. Imaging microglial activation in Huntington's disease. Brain Res Bull. (2007) 72:148-151.

[77] Mastrokolias A, Pool R, Mina E, Hettne KM, van Duijn E et al. Integration of targeted metabolomics and transcriptomics identifies deregulation of phosphatidylcholine metabolism in Huntington's disease peripheral blood samples. Metabolomics. (2016) 12:137.

[78] Lloyd CM, Richardson MP, Brooks DJ, Al-Chalabi A, Leigh PN. Extramotor involvement in ALS: PET studies with GABAA ligand [11C] flumazenil. Brain (2000) 123: 2298-2296.

[79] Turner NR, Cagnin A, Turkheimer FE, Miller CCJ, Shaw CE, et al. Evidence of widespread cerebral microglial activation in amyotrophic lateral sclerosis: an [11C] (R)-PK11195 positron emission tomography study. Neurobiol Dis. (2004) 15: 601-609.

[80] Karitzky J, Ludolph AC. Imaging and neurochemical markers for diagnosis and disease progression in ALS. J Neurol Sci. (2001) 191:35-41.

[81] Vogels OMJ, Veltman J, Oyen WJG, Horstink MWI. Decreased striatal dopamine D2 receptor binding in amyotrophic lateral sclerosis (ALS) and multiple system atrophy (MSA): D2 receptor down-regulation versus striatal cell degeneration. J Neurol Sci. (2000) 180: 62-65.

[82] Petri S, Kollewe K, Grothe C, Hori A, Dengler R, et al. GABAA-receptor mRNA expression in the prefrontal and temporal cortex of ALS patients. J Neurol Sci. (2006) 250: 124-132.

[83] Tumani H, Hartung H-P, Hemmer B, Teunissen C, Deisenhammer F, et al. Cerebrospinal fluid biomarkers in multiple sclerosis. Neurobiol Dis. (2009) 35: 117-127.

[84] Pustovrh I, Predić P, Hrastnik D, Gregorić E. Tc-99m-MIBI brain SPECT in diagnosis of multiple sclerosis acute phase. J Neurol Sci. (1997):(Slupp 1):S329.

[85] Bakshi R, Minagar A, Jaisani Z, Wolinsky JS. Imaging of multiple sclerosis: role in neurotherapeutics. NeuroRx_. Journal of the American Society for Experimental NeuroTherapeutics. (2005) 2: 277-303.

[86] Filippi M, Rocca MA. Present and future of fMRI in multiple sclerosis. Expert Rev Neurother. (2013) 13:27-31.

[87] Mangalam A, Poisson L, Nemutlu E, Datta I, Denic A, et al. Profile of circulatory metabolites in a relapsing-remitting animal model of multiple sclerosis using global metabolomics. J Clin Cell Immunol (2013) 4. 10.4172/2155-9899.1000150.

[88] Apostolova LG, Steiner CA, Akopyan GG, Dutton RA, Hayashi KM, et al. Three-dimensional gray matter atrophy mapping in mild cognitive impairment and mild Alzheimer disease. Arch Neurol. (2007) 64:1489-1495.

[89] Thompson PM, Woods RP, Mega MS, Toga AW. Mathematical/ computational challenges in creating deformable and probabilistic atlases of the human brain. Hum Brain Mapp. (2000) 9:81-92.

[90] Brodmann K. On the comparative localization of the cortex. Springfield, IL: Thomas; 1960. Translated from Brodman, K. Vergleichende Lokalizationslehre der Grosshirnrinde in ihren Prinzipien dargestellt auf Grund des Zellenbaues, 1909, Barth, Leipzig.

[91] Yushkevich PA, Piven J, Hazlett HC, Smith RG, Ho S et al. User-guided 3D active contour segmentation of anatomical structures: significantly improved efficiency and reliability. Neuroimage. (2006) 31:1116-1128.

[92] Pitiot A, Delingette H, Thompson PM, Ayache N. Expert knowledge-guided segmentation system for brain MRI. Neuroimage. (2004) 23:S85–S96.

[93] Hogan RE, Mark KE, Wang L, Joshi S, Miller MI et al. Mesial temporal sclerosis and temporal lobe epilepsy: MR imaging deformation-based segmentation of the hippocampus in five patients. Radiology. (2000) 216:291-297.

[94] Ashburner J, Friston KJ. Why voxel-based morphometry should be used. Neuroimage. (2001) 14:1238-1243.

[95] Friston KJ, Holmes AP, Worsley KJ, Poline J-P, Frith CD, et al. Statistical parametric maps in functional imaging: a general linear approach. Hum Brain Mapp. (1995) 2:189-210.

[96] Good CD, Johnsrude IS, Ashburner J, Henson RN, Friston KJ et al. A voxel-based morphometric study of ageing in 465 normal adult human brains. Neuroimage. (2001) 14:21-36.

[97] Busatto GF, Garrido GE, Almeida OP, Castro CC, Camargo CHP, et al. A voxel-based morphometry study of temporal lobe gray matter reductions in Alzheimer's disease. Neurobiol Aging. (2003) 24: 221-231.

[98] Mummery CJ, Patterson K, Price CJ, Ashburner J, Frackowiak RS et al. A voxel-based morphometry study of semantic dementia: relationship between temporal lobe atrophy and semantic memory. Ann Neurol. (2000) 47:36-45.

[99] Gitelman DR, Ashburner J, Friston KJ, Tyler LK, Price CJ et al. Voxel-based morphometry of herpes simplex encephalitis. Neuroimage. (2001) 13:623-631.

[100] Whitwell JL, Weigand SD, Shiung MM, Boeve BF, Ferman TJ, et al. Focal atrophy in dementia with Lewy bodies on MRI: a distinct pattern from Alzheimer's disease. Brain. (2007) 130:708-719.

[101] Salmond CH, Ashburner J, Vargha-Khadem F, Connelly A, Gadian DG et al. Distributional assumptions in voxel-based morphometry. Neuroimage. (2002) 17:1027-1030.

[102] Gee JC, Thompson PM. Guest editorial: Special issue on computational

neuroanatomy. IEEE Trans Med Imaging. (2007) 26:425-426.

[103] Chiang MC, Dutton RA, Hayashi KM, Lopez OL, Aizenstein HJ, et al. 3D pattern of brain atrophy in HIV/AIDS visualized using tensor-based morphometry. Neuroimage. (2007) 34:44-60.

[104] Thompson PM, Hayashi KM, de Zubicaray G, Janke AL, Rose SE, et al. Dynamics of gray matter loss in Alzheimer's disease. J Neurosci. (2003) 23:994-1005.

[105] Frisoni GB, Pievani M, Testa C, Sabattoli F, Bresciani L, et al. The topography of grey matter involvement in early and late onset Alzheimer's disease. Brain. (2007)130:720-730.

[106] Bookheimer SY, Braskie MN, Burggren AC., et al. Functional MRI and structure-function mapping in the early diagnosis of Alzheimer's disease. Annual Meeting of the Americal College of Neuropsychopharmacology; Hollywood, FL. 2006.

[107] Ballmaier M, O'Brien JT, Burton EJ, Thompson PM, Rex DE, et al. Comparing gray matter loss profiles between dementia with Lewy bodies and Alzheimer's disease using cortical pattern matching: diagnosis and gender effects. Neuroimage. (2004) 23:325-335.

[108] Bäckman L, Jones S, Berger AK, Laukka EJ, Small BJ. Cognitive impairment in preclinical Alzheimer's disease: a meta-analysis. Neuropsychology. (2005)19:520-531.

[109] Agarwal R, Chhillar N, Kushwaha S, Singh NK, Tripathi CB. Role of vitamin B(12), folate, and thyroid stimulating hormone in dementia: A hospital-based study in north Indian population. Ann Indian Acad Neurol. (2010) 13:257-262.

[110] Price BR, Wilcock DM, Weekman EM. Hyperhomocysteinemia as a Risk Factor for Vascular Contributions to Cognitive Impairment and Dementia. Front Aging Neurosci. (2018) 10:350.

[111] Barnes DE, Alexopoulos GS, Lopez OL, Williamson JD, Yaffe K. Depressive symptoms, vascular disease, and mild cognitive impairment: findings from the Cardiovascular Health Study. Arch. Gen. Psychiatry. (2006) 63: 273-279.

[112] Monastero R, Palmer K, Qiu C, Winblad B, Fratiglioni L. Heterogeneity in risk factors for cognitive impairment, no dementia: population-based longitudinal study from the Kungsholmen Project. Am J Geriatr Psychiatry. (2007) 15: 60-69.

[113] Qiu C, Kivipelto M, von Strauss E. Epidemiology of Alzheimer's disease: occurrence, determinants, and strategies toward intervention. Dialogues Clin Neurosci. (2009) 11:111-128.

[114] Poddar MK, Banerjee S, Chakraborty A, Dutta D.Metabolic disorder in Alzheimer's disease. Metab Brain Dis. Communicated. (2020).

[115] Gatz M, Reynolds CA, Fratiglioni L, Johansson B, Mortimer JA, et al. Role of genes and environments for explaining Alzheimer disease. Arch. Gen. Psychiatry. (2006) 63:168-174.

[116] Fratiglioni L, von Strauss E, Qiu C. Epidemiology of the dementias of old age. In: Jacoby R, Oppenheimer C, Dening T, Thomas A. (Eds.), The Oxford Textbook of Old Age Psychiatry. Oxford University Press, New York, (2008): 391-406.

[117] Bertram L, Tanzi RE. The genetic epidemiology of neurodegenerative disease. J. Clin. Invest. (2005) 115: 1449-1457.

[118] Xu W, Qiu C, Winblad B, Fratiglioni L. The effect of borderline diabetes on the risk of dementia and Alzheimer's disease. Diabetes. (2007), 56:211-216.

[119] Kröger E, Andel R, Lindsay J, Benounissa Z, Verreault R, et al. Is complexity of work associated with risk of dementia? The canadian study of health and aging. Am. J. Epidemiol. (2008) 167:820-830.

[120] Alzheimer's disease facts and figures. Alzheimer's Dement. (2020) 16:391-460.

[121] Palmer S, RD. Smart Eating— How Diet May Help Preserve the Brain. (2009) 11(7):24.

[122] Valenzuela R, Valenzuela B. Docosahexaenoic Acid (DHA), in the Prevention and Treatment of Neurodegenerative Diseases. Neurodegenerative Diseases - Processes, Prevention, Protection and Monitoring, Raymond Chuen-Chung Chang, IntechOpen. (2011) 10.5772/28000.

[123] Siddiqui R, Harvey K, Bammerlin E, Ikhlaque N. Docosahexaenoic Acid: A Potential Modulator of Brain Tumors and Metastasis. J. Biomol. Res. Ther. (2013) 2:10.4172/2167-7956.1000e119.

[124] Gomez-Pinilla F, Nguyen TT. Natural mood foods: the actions of polyphenols against psychiatric and cognitive disorders. Nutr Neurosci. (2012) 15:127-133.

[125] Di Meo F, Margarucci S, Galderisi U, Crispi S, Peluso G. Curcumin, Gut Microbiota, and Neuroprotection. Nutrients. (2019) 11:2426.

[126] Bourre JM. Effects of nutrients (in food) on the structure and function of the nervous system: update on dietary requirements for brain. Part 1: micronutrients. J Nut Health Aging (2006) 10:377-385.

[127] Luchsinger JA, Tang MX, Miller J, Green R, Mayeux R. Relation of higher folate intake to lower risk of Alzheimer disease in the elderly. Arch Neurol. (2007) 64:86-92.

[128] Gu Y, Schupf N, Cosentino SA, Luchsinger JA, Scarmeas N. Nutrient intake and plasma β-amyloid. Neurology. (2012) 78:1832-1840.

[129] Fratiglioni L, Winblad B, von Strauss E. Prevention of Alzheimer's disease and dementia. Major findings from the Kungsholmen Project. Physiol Behav. (2007) 92:98-104.

[130] Berti V, Murray J, Davies M, Spector N, Tsui WH, et al. Nutrient patterns and brain biomarkers of Alzheimer's disease in cognitively normal individuals. J Nutr Health Aging. (2015) 19:413-423.

[131] Chakraborty A, Banerjee S, Mukherjee B, Poddar MK. Calorie restriction improves aging-induced impairment of cognitive function in relation to deregulation of brain regional GABA system and corticosterone status. Mech Ageing Dev (2020) 189: 111242. 248.

[132] Zhang Y, Liu C, Zhao Y, Zhang X, Li B, et al. The Effects of Calorie Restriction in Depression and Potential Mechanisms. Curr Neuropharmacol. (2015) 13:536-542.

[133] Ashrafian H, Harling L, Darzi A, Athanasiou T. Neurodegenerative disease and obesity: what is the role of weight loss and bariatric interventions? Metab Brain Dis. (2013) 28:341-353.

[134] Dudar JD, Whishaw IQ, Szerb JC. Release of acetylcholine from the hippocampus of freely moving rats during sensory stimulation and

running. Neuropharmacology. (1979) 18:673-678.

[135] Maass A, Duzel S, Brigadski T, Goerke M, Becke A, et al. Relationships of peripheral IGF-1, VEGF and BDNF levels to exercise-related changes in memory, hippocampal perfusion and volumes in older adults. NeuroImage. (2016) 131:142-154.

[136] Sleiman SF, Henry J, Al-Haddad R, El Hayek L, Abou Haidar E, et al. Exercise promotes the expression of brain derived neurotrophic factor (BDNF) through the action of the ketone body beta-hydroxybutyrate. Elife. (2016) 5:e15092.

[137] Ieraci A, Madaio AI, Mallei A, Lee FS, Popoli M. Brain-derived neurotrophic factor Val66Met human polymorphism impairs the beneficial exercise-induced neurobiological changes in mice. Neuropsychopharmacology. (2016) 41:3070-3079.

[138] Wrann CD, White JP, Salogiannnis J, Laznik-Bogoslavski D, Wu J, et al. Exercise induces hippocampal BDNF through a PGC-1alpha/FNDC5 pathway. Cell Metab. (2013) 18:649-659.

[139] Allard JS, Ntekim O, Johnson SP, Ngwa JS, Bond V, et al. APOEε4 impacts up-regulation of brain-derived neurotrophic factor after a six-month stretch and aerobic exercise intervention in mild cognitively impaired elderly African Americans: A pilot study. Exp Gerontol. (2017) 87:129-136.

[140] Khorshid AT, Acosta C, Cortes C, Lakowski T, Gangadaran S, et al. Transcriptional regulation of brain-derived neurotrophic factor (BDNF) by methyl CpG binding protein 2 (MeCP2): a novel mechanism for re-myelination and/or myelin repair involved in the treatment of multiple sclerosis (MS). Mol Neurobiol. (2016) 53:1092-1107.

[141] Gao Y, Zhao Y, Pan J, Yang L, Huang T, et al. Treadmill exercise promotes angiogenesis in the ischemic penumbra of rat brains through caveolin-1/VEGF signaling pathways. Brain Res. (2014) 1585:83-90.

[142] Rich B, Scadeng M, Yamaguchi M, Wagner PD, Breen EC. Skeletal myofiber vascular endothelial growth factor is required for the exercise training-induced increase in dentate gyrus neuronal precursor cells. J Physiol. (2017) 595:5931-5943.

[143] Cotman CW, Berchtold NC, Christie LA. Exercise builds brain health: key roles of growth factor cascades and inflammation. Trends Neurosci. (2007) 30:464-472.

[144] Knochel C, Oertel-Knochel V, O'Dwyer L, Prvulovic D, Alves G, et al. Cognitive and behavioural effects of physical exercise in psychiatric patients. Prog Neurobiol. (2012) 96:46-68.

[145] Zimmer P, Stritt C, Bloch W, Schmidt FP, Hubner ST, et al. The effects of different aerobic exercise intensities on serum serotonin concentrations and their association with Stroop task performance: a randomized controlled trial. Eur J Appl Physiol. (2016) 116:2025-2034.

[146] Audhya T, Adams JB, Johansen L. Correlation of serotonin levels in CSF, platelets, plasma, and urine. Biochim Biophys Acta. (2012) 1820:1496-1501.

[147] Law LL, Rol RN, Schultz SA, Dougherty RJ, Edwards DF, et al. Moderate intensity physical activity associates with CSF biomarkers in a cohort at risk for Alzheimer's disease. Alzheimers Dement (Amst). (2018) 10:188-195.

[148] Schenkman M, Moore CG, Kohrt WM, Hall DA, Delitto A, et al. Effect of high-intensity treadmill exercise on motor symptoms in patients

with de novo Parkinson Disease: a phase 2 randomized clinical trial. JAMA Neurol. (2018) 75:219-226.

[149] Shin MS, Kim TW, Lee JM, Ji ES, Lim BV. Treadmill exercise alleviates nigrostriatal dopaminergic loss of neurons and fibers in rotenone-induced Parkinson rats. J Exerc Rehabil. (2017) 13:30-35.

[150] Shaw LM, Korecka M, Clark CM, Lee V M-Y, Trojanowski JQ. Biomarkers of neurodegeneration for diagnosis and monitoring therapeutics. Nat Rev Drug Discov. (2007) 6:295-303.

[151] Tomita T, Iwatsubo T. The inhibition of gamma-secretase as a therapeutic approach to Alzheimer's disease. Drug News Perspect. (2004)17:321-325.

[152] Arosio P, Vendruscolo M, Dobson CM, Knowles TP. Chemical kinetics for drug discovery to combat protein aggregation diseases. Trends Pharmacol Sci. (2014) 35:127-135.

[153] Lee JH, Jeong SK, Kim B, Park K, Dash A. Donepezil across the spectrum of Alzheimer's disease: Dose optimization and clinical relevance. Acta Neurol Scand. (2015) 131:259-267.

[154] Bonuccelli U, Colzi A, Del Dotto P. Pergolide in the treatment of patients with early and advanced Parkinson's disease. Clin. Neuropharmacol. (2002) 25:1-10.

[155] McMurray CT. Huntington's disease: New hope for therapeutics. Trends Neurosci. (2001) 24: S32–S38.

[156] Faissner S, Gold R. Oral therapies for multiple sclerosis. Cold Spring Harb. Perspect. Med. (2019) 9:a032011.

[157] Hussain R, Zubair H, Pursell S, Shahab M Neurodegenerative Diseases: Regenerative Mechanisms and Novel Therapeutic Approaches. Brain Sciences. 2018 Sep;8(9). DOI: 10.3390/brainsci8090177.

[158] Chaturvedi SK, Alam P, Khan JM, Siddiqui MK, Kalaiarasan P, et al. Biophysical insight into the anti-amyloidogenic behavior of taurine. Int J Boil Macromol. (2015) 80: 375-384.

[159] Hartl FU. Protein misfolding diseases. Annu Rev Biochem. (2017) 86: 21-26.

[160] Jackrel ME, Shorter J. Engineering enhanced protein disaggregases for neurodegenerative disease. Prion. (2015) 9: 90-109.

[161] Patterson KR, Ward SM, Combs B, Voss K, Kanaan NM et al. Heat shock protein 70 prevents both tau aggregation and the inhibitory effects of preexisting tau aggregates on fast axonal transport. Biochemistry. (2011) 50: 10300-10310.

[162] Evans CG, Wisén S, Gestwicki JE. Heat shock proteins 70 and 90 inhibit early stages of amyloid beta (1-42) aggregation in vitro. J Boil Chem (2006) 281(44):33182-33191.

[163] Maiti P, Manna J, Veleri S, Frautschy S. Molecular chaperone dysfunction in neurodegenerative diseases and effects of curcumin. BioMed Res Int. (2014) 2014: 495091.

[164] Dedmon MM, Christodoulou J, Wilson MR, Dobson CM. Heat shock protein 70 inhibits alpha-synuclein fibril formation via preferential binding to prefibrillar species. J Boil Chem (2005) 280: 14733-14740.

[165] Ciechanover A, Kwon YT. Protein quality control by molecular chaperones in neurodegeneration. Front. Neurosci. (2017) 11:185.

[166] Björkqvist M, Wild EJ, Thiele J, Silvestroni A, Andre R, et al. A novel pathogenic pathway of immune

activation detectable before clinical onset in Huntington's disease. J. Exp. Med. (2008) 205:1869-1877.

[167] Miron VE, Boyd A, Zhao JW, Yuen TJ, Ruckh JM et al. M2 microglia and macrophages drive oligodendrocyte differentiation during CNS remyelination. Nat. Neurosci. (2013) 16: 1211-1218.

[168] Träger U, Andre R, Lahiri N, Magnusson-Lind A, Weiss A, et al. HTT-lowering reverses Huntington's disease immune dysfunction caused by NF_B pathway dysregulation. Brain J Neurol. (2014) 137:819-833.

[169] Brochard V, Combadiere B, Prigent A, Laouar Y, Perrin A, et al. Infiltration of CD4+ lymphocytes into the brain contributes to neurodegeneration in a mouse model of Parkinson disease. J Clin Investig. (2009) 119:182-192.

[170] Games D, Valera E, Spencer B, Rockenstein E, Mante M, et al. Reducing C-terminal-truncated alpha-synuclein by immunotherapy attenuates neurodegeneration and propagation in Parkinson's disease-like models. J. Neurosci. (2014) 34: 9441-9454.

[171] Boland B, Kumar A, Lee S, Platt FM, Wegiel J, et al. Autophagy induction and autophagosome clearance in neurons: Relationship to autophagic pathology in Alzheimer's disease. J Neurosci. (2008) 28: 6926-6937.

[172] Nixon RA. The role of autophagy in neurodegenerative disease. Nat Med. (2013) 19: 983-997.

[173] Pupyshev AB, Korolenko TA, Tikhonova MA. A therapeutic target for inhibition of neurodegeneration: Autophagy. Neurosci Behav Physiol. (2017) 47:1109-1127.

[174] Hara T, Nakamura K, Matsui M, Yamamoto A, Nakahara Y, et al. Suppression of basal autophagy in neural cells causes neurodegenerative disease in mice. Nature. (2006) 441: 885-889.

[175] Liang C-C, Wang C, Peng X, Gan B, Guan J-L. Neural-specific deletion of FIP200 leads to cerebellar degeneration caused by increased neuronal death and axon degeneration. J Boil Chem. (2010) 285: 3499-3509.

[176] Choi SJ, Panhelainen A, Schmitz Y, Larsen KE, Kanter E, et al. Changes in neuronal dopamine homeostasis following 1-methyl-4-phenylpyridinium (MPP+) exposure. J Boil Chem. (2015) 290: 6799-6809.

[177] Steele JW, Gandy S. Latrepirdine (Dimebon®), a potential alzheimer therapeutic, regulates autophagy and neuropathology in an Alzheimer mouse model. Autophagy. (2013) 9: 617-618.

[178] Berger Z, Ravikumar B, Menzies FM, Oroz LG, Underwood BR, et al. Rapamycin alleviates toxicity of different aggregate-prone proteins. Hum Mol Genet. (2005) 15: 433-442.

[179] Sun Q, Wei L-L, Zhang M, Li T-X, Yang C, et al. Rapamycin inhibits activation of ampk-mTOR signaling pathway-induced Alzheimer's disease lesion in hippocampus of rats with type 2 diabetes mellitus. Int J Neurosci. (2018) 2018:1-22.

[180] Allen SJ, Watson JJ, Dawbarn D. The neurotrophins and their role in Alzheimer's disease. Curr Neuropharmacol. (2011) 9:559-573.

[181] Jiao SS, Shen LL, Zhu C, Bu XL, Liu YH, et al. Brain-derived neurotrophic factor protects against tau-related neurodegeneration of Alzheimer's disease. Transl Psychiatry. (2016) 6: e907.

[182] Middlemas DS, Kihl BK, Zhou J, Zhu X. Brain-derived neurotrophic

factor promotes survival and chemoprotection of human neuroblastoma cells. J Boil Chem. (1999) 274: 16451-16460.

[183] Weissmiller AM, Wu C. Current advances in using neurotrophic factors to treat neurodegenerative disorders. Transl Neurodegener. (2012) 1:14.

[184] Costa A, Peppe A, Carlesimo GA, Zabberoni S, Scalici F, et al. Brain-derived neurotrophic factor serum levels correlate with cognitive performance in Parkinson's disease patients with mild cognitive impairment. Front Behav Neurosci. (2015) 9:253.

[185] Pan W, Banks WA, Kastin AJ. Permeability of the blood–brain barrier to neurotrophins. Brain Res. (1998) 788: 87-94.

[186] Markianos M, Panas M, Kalfakis N, Vassilopoulos D. Plasma testosterone in male patients with Huntington's disease: Relations to severity of illness and dementia. Ann Neurol. (2005) 57:520-525.

[187] Bove R, Musallam A, Healy BC, Raghavan K, Glanz BI, et al. Low testosterone is associated with disability in men with multiple sclerosis. Mult Scler. (2014) 20:1584-1592.

[188] Dye RV, Miller KJ, Singer EJ, Levine AJ. Hormone replacement therapy and risk for neurodegenerative diseases. Int J Alzheimers Dis. (2012) 2012:258454.

[189] Blázquez E, Velázquez E, Hurtado-Carneiro V, Ruiz-Albusac JM. Insulin in the brain: Its pathophysiological implications for states related with central insulin resistance, type 2 diabetes and Alzheimer's disease. Front Endocrinol. (2014) 5:161.

[190] Talbot K, Wang H-Y, Kazi H, Han L-Y, Bakshi KP, et al. Demonstrated brain insulin resistance in Alzheimer's disease patients is associated with IGF-1 resistance, IRS-1 dysregulation, and cognitive decline. J Clin Investig. (2012) 122:1316-1338.

[191] Kang H, Khang R, Ham S, Jeong GR, Kim H, et al. Activation of the ATF2/CREB-PGC-1_ pathway by metformin leads to dopaminergic neuroprotection. Oncotarget. (2017) 8:48603-48618.

[192] Bowen DM, Smith CB, White P, Davison AN. Neurotransmitter-related enzymes and indices of hypoxia in senile dementia and other abiotrophies. Brain J Neurol. (1976) 99: 459-496.

[193] Whitehouse PJ, Price DL, Clark AW, Coyle JT, DeLong MR. Alzheimer disease: Evidence for selective loss of cholinergic neurons in the nucleus basalis. Ann Neurol Off J Am Neurol Assoc Child Neurol Soc. (1981) 10:122-126.

[194] Drachman DA, Leavitt J. Human memory and the cholinergic system: A relationship to aging? Arch Neurol. (1974) 30:113-121.

[195] Massoud F, Gauthier S. Update on the pharmacological treatment of Alzheimer's disease. Curr Neuropharmacol. (2010) 8:69-80.

[196] Barua S, Kim JY, Yenari MA, Lee JE. The role of NOX inhibitors in neurodegenerative diseases. IBRO Reports (2019) 7: 59-69.

[197] Pal R, Cristan EA, Schnittker K, Narayqn M. Rescue of ER oxidoreductase function through polyphenolic phytochemical intervention: Implications for subcellular traffic and neurodegenerative disorders. Biochem Biophys Res Commun (2010) 392: 567-571.

[198] Gopinath K, Sudhandiran G. Naringin modulates oxidative stress and

inflammation in 3-nitropropionic acid-induced neurodegeneration through the activation of nuclear factor-erythroid-2-related factor-2 signalling pathway. Neurosci (2012) 227: 134-143.

[199] Banerjee S. Aging-induced neurodegeneration in relation to brain regional A-beta deposition, locomotor and cognitive function: role of carnosine. J Neurochem (2019) 150: 125.

[200] Banerjee S, Mukherjee B, Poddar MK. Carnosine restores aging-induced elevation of corticosterone status and brain regional amyloid-beta in relation to down regulation of locomotor activity. J Syst Integr Neurosci (2019) 5: 1-11.

[201] Karim MF, Banerjee S, Poddar MK. Does patchouli oil change blood platelet monoamine oxidase-A activity of adult mammals? J Physiol Sci (2018) 68: 281-291.

[202] Kilgore M, Miller CA, Fass DM, Hennig KM, Haggarty SJ, et al. Inhibitors of class 1 histone deacetylases reverse contextual memory deficits in a mouse model of Alzheimer's disease. Neuropsychopharmacology. (2010) 35(4):870-880.

[203] Jowaed A, Schmitt I, Kaut O, Wullner U. Methylation regulates alpha-synuclein expression and is decreased in Parkinson's disease patients' brains. J Neurosci. (2010) 30:6355-6359.

[204] Outeiro TF, Kontopoulos E, Altmann SM, Kufareva I, Strathearn KE, et al. Sirtuin 2 inhibitors rescue alpha-synuclein-mediated toxicity in models of Parkinson's disease. Science. (2007) 317:516-519.

[205] Nicholas AP, Lubin FD, Hallett PJ, Vattem P, Ravenscroft P, et al. Striatal histone modifications in models of levodopa-induced dyskinesia. J Neurochem. (2008) 106:486-494.